Lecture Notes in Mathematics

A collection of informal reports and seminars
Edited by A. Dold, Heidelberg and B. Eckmann, Zürich

Series: Mathematisches Institut der Universität Heidelberg
Adviser: K. Krickeberg

328
Decidable Theories II

Edited by G. H. Müller and D. Siefkes

J. Richard Büchi

Purdue University, Lafayette, IN/USA

Dirk Siefkes

Mathematisches Institut der Universität Heidelberg,
Heidelberg/BRD

The Monadic Second Order Theory of All Countable Ordinals

Springer-Verlag
Berlin · Heidelberg · New York 1973

AMS Subject Classifications (1970): 02-02, 02F10, 02G05, 02G10, 68-02, 68A25, 02G20, 02H10, 02K20

ISBN 3-540-06345-5 Springer-Verlag Berlin · Heidelberg · New York
ISBN 0-387-06345-5 Springer-Verlag New York · Heidelberg · Berlin

Offsetdruck: Julius Beltz, Hemsbach/Bergstr.

Table of Contents

THE MONADIC SECOND ORDER THEORY OF ω_1

J. Richard Büchi

We will present here (Section 6) a method for deciding truth of monadic second-order sentences (Section 1) in the system $[\omega_1,<]$, the order relation on the set ω_1 of all countable ordinal numbers. If the ordinal α is identified with the set of all ordinals $\xi<\alpha$ (von Neumann), the relation $<$ becomes the membership relation $\in$, and ω_1 becomes the first uncountable ordinal. Let $\mathcal{P}x$ denote the set of all subsets of x. Our result may be restated in the form: The first-order theory of $[\mathcal{P}\omega_1,\in]$ is decidable.

The axiom of choice will be used at various places. If σ and τ are cardinals let AC_σ^τ be the axiom of choice from $\leq\sigma$ sets, each of cardinality $\leq\tau$. In $MT[\alpha,<]$, the monadic (second-order) theory of the ordinal α, it is easy to make up a sentence Σ which states "α is ω_0-accessible". (In fact, the presence of ω_n-accessibility statements is an important feature of the MT of linear orders.) Now, the statement "ω_1 is ω_0-accessible" depends on whether or not $AC_{\omega_0}^{\omega_0}$. Thus, the same decision-method cannot work for $MT[\omega_1,<]$, both in case $AC_{\omega_0}^{\omega_0}$, and in case $\sim AC_{\omega_0}^{\omega_0}$. Our method works in case $AC_{\omega_1}^{\gamma}$, where γ is the cardinality of the continuum. In fact, this stronger part of AC will be used in two places. One of these uses seems

This work was supported by Grant no. GJ-980 from the N.S.F.

essential; it goes as follows.

Let $\mathcal{J}_1$ be the filter of cofinal closed subsets of ω_1. Using a method of Ulam [17], which depends on $AC_{\omega_1}^{\gamma}$, we will show in Section 5 : The Boolean algebra $\mathcal{P}\omega_1/\mathcal{J}_1$ has no atom. This fact will be used to show that our decision-method works for $MT[\omega_1,<]$. It is interesting to note that the assertion "$\mathcal{P}\omega_1/\mathcal{J}_1$ has no atom" can be stated as a sentence Σ in $MT[\omega_1,<]$, and the same goes for "$\mathcal{J}_1$ is a filter" (depends on $AC_{\omega_0}^{\omega_0}$, or "ω_1 is not ω_0-accessible"), and "$\mathcal{J}_1$ is not prime" (depends on Ulam, and "$\mathcal{J}_1$ is a ω_0-filter"). In fact, the presence of assertions about such filters as $\mathcal{J}_1$, is a second important feature of the MT of linear orders. This, and the matter of accessibility, should be minded by one who tries his hand on the important unsolved problems on MT of linear orders.

In [3] we outlined a decision-method for the monadic second order theory $MT[\alpha,<]$, for any countable ordinal α. Details of this proof are essential for the understanding of $MT[\omega_1,<]$, we present it in Section 4. In a paper with D. Siefkes we will show that the basic results of Section 4 answer all questions concerning the theory T of all countable well-orders: T is decidable, and recursively axiomatizable, on the base of the usual rules of proof for (two-sorted) elementary logic. We will rather concretely describe the Tarski-Lindenbaumm algebra over T, and survey the complete extensions of T, including non-standard ones. Also matters of definability in T will be discussed.

The proofs presented in this paper afford further examples of what seems to be a quite general method for eliminating quantifiers in monadic second order theories. All known cases of decidable MT-theories can be obtained by using this method. In Section 2 we outline the method, survey the results, and state outstanding problems, which seem accessible by our method. In Section 3 we show how the two versions of the method work in the classical case $MT[\omega,<]$. To implement the deterministic method, we present the important subset-construction of McNaughton [13], in a streamlined and more explicit version. In this form the construction will naturally extend to arbitrary ordinals $\alpha<\omega_1$; see Section 4. The non-deterministic method was first applied in [2], where a combinatorial theorem of Ramsey's was used. We will see that Ramsey is not really needed, and even the infinity lemma (König) is not used. This makes the non-deterministic method available at ω_1, see Section 6.

Acknowledgment: D.Siefkes has found that (besides in accessibility matters) AC is used in the "splicing principle". This use of AC will be discused in the paper [BS] , the second paper in this volume. I thank D.Siefkes and Ch.Zaiontz for critical reading of the manuscript.

1. Monadic second order theories; the prenex form lemma

Monadic second-order formulas (MT-formulas) are built up from letters or primitives (such as the individual letter 0, the function letter ', the predicate or relation letter $<$) by using propositional connectives ($\wedge$, $\vee$, $\sim$, $\supset$, $\equiv$, T, F), individual variables ($x, y, z, t, \ldots\ldots$), monadic predicate variables or set-variables ($X, Y, Z, \ldots\ldots$) and the quantifiers $\forall$, $\exists$, used with respect to both types of variables. Also notations such as Yx occur in MT-formulas, and $x \in Y$ might have been used instead. Thus, $\in$ should actually be listed among the primitives occurring in MT-formulas, and it makes good sense to allow the use of other letters which are of such types as "set of sets", "relation between sets and individuals", etc. An MT-sentence is an MT-formula in which every occurrence of a variable is bound by a quantifier. Of course, we have in mind precisely formulated formation-rules which tell how MT-formulas must be built up. In place of stating these rules we give some interesting examples of MT-formulas and sentences.

Ind: $(\forall X).[X0 \wedge (\forall t)[Xt \supset Xt']] \supset (\forall t)Xt$

Wlo: $(\forall X).(\exists t)Xt \supset (\exists t)[Xt \wedge (\forall y)[y<t \supset \sim Xy]]$

Sml(x,y): $(\forall Z).[Zx' \wedge (\forall t)[Zt \supset Zt']] \supset Zy$

Scs(x,y): $x<y \wedge \sim(\exists t)[x<t \wedge t<y]$

x=y: $(\forall Z).Zx \supset Zy$ $\qquad$ x≦y: $x=y \vee x<y$

X⊆Y: $(\forall t).Xt \supset Yt$ $\qquad$ X=Y: $X \subseteq Y \wedge Y \subseteq X$

The sentence Ind may have been denoted more explicitly by Ind(0,'), indicating that no primitives, except 0 and ', occur. By using the notation Sml(x,y) we indicate that no variables, except x and y, have free occurrences in this formula. We will use notations like $\mathcal{A}(X,Y)$, $\mathcal{C}$ to range over formulas and sentences (formulas without free variables).

A (mathematical) structure $\underline{D} = [D, R_1,\dots,R_n]$ consists of a set D, called its domain, and objects $R_1,\dots,R_n$ belonging to the echelon over D (such as elements of D, operators and relations on D, and possibly higher type objects such as relation on subsets of D). MT-formulas Σ may be interpreted in a structure $\underline{D}$, which to any primitive letter occurring in Σ assigns an object R of $\underline{D}$. This assignment must be such that the type of the object fits that of the corresponding letter. To make clear which assignment is meant, in a particular case, we will ambiguously use the same notation for the letter and the corresponding object.

The basic semantic notion, which relates sentences Σ to structures $\underline{D}$, is the model-relation $\Sigma \succ\!\!- \underline{D}$, read "$\Sigma$ holds in $\underline{D}$", or "$\underline{D}$ is a model of Σ". For example we have, Ind $\succ\!\!-$ $[\omega,0,']$, which just says that the induction axiom holds in the natural number system. If $[D,<]$ is a linear order, then Wlo $\succ\!\!-$ $[D,<]$ asserts that $[D,<]$ is a well-order. Again, we will not state a rigorous definition (see Tarski) of the model-relation for MT-sentences. The intuitive meaning of "Σ holds in $\underline{D}$" or "$\underline{D}$ satisfies the axiom Σ" seems sufficiently clear for most mathematical discussions; it will serve well in the present context. However, note that, among other things, $\Sigma \succ\!\!- \underline{D}$ implies that all letters in Σ correspond one-to-one to some (or all) objects of $\underline{D}$, that individual variables are interpreted to range over D, and that set-variables are interpreted to range over $\mathcal{P}D$ = set of all subsets of D.

Closely related to the model-relation is the concept "the formula Φ defines the relation R in $\underline{D}$". Here, Φ may contain free variables (say x and Y) and R is an object in the echelon over D, whose type corresponds to the free variables in Φ (in the example, $R \subseteq D \times \mathcal{P}D$). For example, in $[\omega,0,']$ the formula $Sml(x,y)$ defines the natural order-relation $x<y$ on ω. In any structure $\underline{D}$, the formula $(\forall t)[Xt \supset Yt]$ defines the inclusion relation $X \subseteq Y$, on $\mathcal{P}D$.

The monadic (second-order) theory of the structure $\underline{D}$ is the set $MT[\underline{D}] = \{\Sigma; \Sigma \succ\!\!- \underline{D}\}$, consisting of all MT-sentences Σ which (fit the type of $\underline{D}$ and) are true in $\underline{D}$. More generally, the monadic theory of a set K of structures (all of the same type) is the set $MT[K] = \cap\{MT[\underline{D}]; \underline{D} \in K\}$, consisting of all MT-sentences which are true in all structures $\underline{D} \in K$. Note that $MT[K] \subseteq MT[\underline{D}]$, for $\underline{D} \in K$. A theory $MT[\underline{D}]$ of one structure is complete in the sense, $\Sigma \in MT[\underline{D}]$ or $\sim\Sigma \in MT[\underline{D}]$, for all MT-sentences Σ.

Let $\underline{D}_1$ and $\underline{D}_2$ denote structures of equal type. Then either $MT[\underline{D}_1] = MT[\underline{D}_2]$, or else $MT[\underline{D}_1, \underline{D}_2] \subset MT[\underline{D}_1]$. In the first case, where the two structures are indistinguishable by MT-sentences, we say that *$\underline{D}_1$ and $\underline{D}_2$ are MT-equivalent*. A structure $\underline{D}$ will be called MT-*categorical* if every MT-equivalent structure is isomorphic to $\underline{D}$. *The theory* MT[K] *is (semi) decidable*, if there is an algorithm, which applies to every MT-sentence Σ, and tells (whether) whether or not $\Sigma \in MT[K]$. *The decision problem* of MT[K] is to find such an algorithm, or to show that none exists.

The operator MT from structures to sentences has a dual SM, from sentences to structures (see discussion of polarities, G. Birkhoff, *Lattice Theory*). Namely, $SM[\Sigma] = \{\underline{D}; \Sigma \succ\!- \underline{D}\}$, *the set of all models of the axiom* Σ. Again, for a class A of sentences, $SM[A] = \cap\{SM[\Sigma]; \Sigma \in A\}$ is the set of all models of the axiom system A. (Such operators were first studied extensively, in the case of first-order sentences, by A. Tarski.) Clearly this is a very basic notion of mathematics; $K = SM[A]$ just means that A *is an axiom system for the set* K *of structures*. If all structures in SM[A] are isomorphic, we say that A is a *categorical axiom system*. If K admits a finite system of MT-axioms we say that K is *finitely* MT-*axiomatizable*.

Let $\theta_0 D$ denote the set of all finite subsets of D, and in interpretations $\underline{D}$ of MT-formulas, let us restrict the range of set-variables to $\mathcal{P}_0 D$. Thus $(\forall X)$ now stands for "for all finite subsets X of D". This leads to a new model-relation $\Sigma \succ\!\!-_0 \underline{D}$ (read Σ holds in the WM-sense in $\underline{D}$). The <u>weak monadic theory</u> of $\underline{D}$ is the set $WMT[\underline{D}] = \{\Sigma; \Sigma \succ\!\!-_0 \underline{D}\}$, consisting of all MT-sentences Σ which hold in $\underline{D}$ in the WM-sense. Similarly $WMT[K]$ of a class K of models is defined, and two structures are WMT- <u>equivalent</u> if their WMT's are equal. The reader might want to show that a structure $[D,a,f]$, which is WMT-equivalent to the number system $[\omega,0,']$, must be isomorphic to it. I.e., the induction axiom is not really needed in a categorical axiom system for the successor function on natural numbers.

Often, in $MT[\underline{D}]$ one can define the set $\mathcal{P}_0 D$ by an MT-formula $Fin(X)$. In such cases $WMT[\underline{D}]$ really is weaker (no stronger) than $MT[\underline{D}]$, in the sense that the former theory can be <u>translated into</u> the latter. Namely, given an MT-sentence Σ, construct the sentence Σ_{Fin} by relativizing all set-quantifiers in Σ to Fin (replace $(\forall X)$--- by $(\forall X)[Fin(X) \supset ---]$ and $(\exists X)$--- by $(\exists X)[Fin(X) \wedge ---]$). Clearly, $\Sigma \in WMT[\underline{D}]$ if and only if, $\Sigma_{Fin} \in MT[\underline{D}]$. The reader will make up an MT-formula $Fin(X)$ which works for any linear order $[D,<]$ (X is finite just in case it is well-ordered and anti-well-ordered). Using the infinity lemma, leads to a formula $Fin(X)$ for the two-splitting tree $[\omega,0,2x+1,2x+2]$ consisting of the root 0 and two free functions which to every vertex x assign the two vertices into which x branches.

As a preliminary to our investigation of monadic second-order theories we will now show that individual quantifiers in MT-formulas may be replaced by set-quantifiers. The idea is simply to replace the individual x by the singleton $\{x\}$.

<u>Lemma</u> 1.1 (<u>the prenex form</u>): To every MT-formula Σ one can find an equivalent formula of form
(predicate prefix).$(\forall x_1 \ldots x_n) M \wedge (\exists t) Q_1 \wedge \ldots \wedge (\exists t) Q_m$. Here $M, Q_1, \ldots, Q_m$ are matrices (i.e., contain no quantifiers), and furthermore each Q_i is of form Xt, whereby X occurs in the prefix.

<u>Proof</u>: Let $(\dot{\exists} X)\Delta$ stand for $(\exists X)[\Delta \wedge (\exists t)Xt]$, and let $(\dot{\forall} X)\Delta$ stand for $(\forall X)[\Delta \vee (\forall t){\sim}Xt]$. We then clearly have the following equivalences:

$$(\dot{\forall} X)(\text{prefix})\Delta \;.\!\equiv\!.\; (\forall X)(\text{prefix})(\forall t)[\Delta \vee {\sim}Xt]$$

(1)

$$(\dot{\exists} X)(\text{prefix})\Delta \;.\!\equiv\!.\; (\exists X)(\text{prefix})[\Delta \wedge (\exists t)Xt]$$

if X does not occur in (prefix), and t does not occur in Δ. Furthermore, from the definition of $\dot{\exists}$ and $\dot{\forall}$, we have:

$$(\forall x)\Delta \;.\!\equiv\!.\; (\dot{\forall} X)(\exists x)[\Delta \wedge Xx]$$

if X does not occur in Δ .

$$(\exists x)\Delta \;.\!\equiv\!.\; (\dot{\exists} X)(\forall x)[\Delta \vee {\sim}Xx]$$

From these one sees,

$$(\forall x)(\text{prefix})\Delta \quad .\!\equiv\!. \quad (\dot{\forall}X)(\text{prefix})(\exists x)[\Delta \wedge Xx]$$

(2)

$$(\exists x)(\text{prefix})\Delta \quad .\!\equiv\!. \quad (\dot{\exists}X)(\text{prefix})(\forall x)[\Delta \vee \sim Xx]$$

if X does not occur in Δ, and x does not occur in (prefix).

We now produce the required formula in the following steps.

Step 1: Find a prenex form $(\text{prefix}_1)M_1$ of Σ.
Step 2: Use the first equivalence (2) to replace all occurrences of universal individual quantifiers in (prefix_1) by quantifiers $\dot{\forall}$. The result is a formula $(\text{prefix}_2)M_2$, such that no $\forall x$ occurs in (prefix_2).
Step 3: Use the second equivalence (2) to replace all individual existential quantifiers in (prefix_2) by quantifiers $\dot{\exists}$. The result is a formula $(\text{prefix}_3)(\forall x_1 \ldots x_s)M_3$, whereby no individual quantifiers occur in (prefix_3).
Step 4: Use the first equivalence (1) to replace all occurrences of $\dot{\forall}$ in (prefix_3) by $\forall$. The result is a formula $(\text{prefix}_4)(\forall x_1 \ldots x_n)M_4$, whereby no individual quantifier, and no $\dot{\forall}$, occurs in (prefix_4).
Step 5: Use the second equivalence (1) to replace all occurrences of $\dot{\exists}$ in (prefix_4) by $\exists$. The result is a formula as required in the lemma. Q.E.D.

Depending on the type of the primitive letters occurring in Σ, the prenex form can be further improved. For example, if all primitives are unary in individuals (such as unary predicates and functions from individuals to individuals, but also binary predicates on sets), the quantifiers $(\forall x_1 \ldots x_n)$ may be moved into the matrix M. The result is a formula in which only one individual variable occurs bound. The lesson for one who studies MT's is: choose your primitive notions carefully; there are many more natural options than in first order theory! For example, in place of $<$ one might choose the class T of all terminal segments as a primitive notion for the MT of linear orders. We will instead use bounded quantifiers to obtain improved prenex forms for MT of linear orders. The following notations will be used throughout this paper.

$(\exists t)_x^y \Delta \ : \ (\exists t)[x \leqq t < y \wedge \Delta]$ "$(\exists t)_x$, $(\exists t)^y$, $(\forall t)_x$, $(\forall t)^y$ are similarly defined."

$(\forall t)_x^y \Delta \ : \ (\forall t)[x \leqq t < y \supset \Delta]$

$(\exists^x t)\Delta \ : \ (\forall y)^x (\exists t)_y^x \Delta$ "Δ holds cofinal x"

$(\forall^x t)\Delta \ : \ (\exists y)^x (\forall t)_y^x \Delta$ "Δ holds terminal x".

X, Y, Z,... often are used to denote n-tuples of set-variables. Similarly, a notation like $H[Xt, Zt]$ may stand for an n-tuple of matrices $H_i[X_1t, \ldots, X_nt, Z_1t, \ldots, Z_mt]$. In this connection we will use abbreviations of the following kind:

$$Xt = a \quad : \quad \bigwedge_{1\leq i\leq n} X_i t \equiv a_i$$ here a stands for an element of $\{F, T\}^n$

$$Zt' = H[Xt,Zt] \quad : \quad \bigwedge_{1\leq i\leq m} Z_i t' \equiv H_i[Xt,Zt]$$

If $X = \langle X_1,\ldots, X_n\rangle$ then the elements of $\{F,T\}^n$ will be called <u>the states of</u> X. In the context of a linear order $[D,<]$ it is useful to think of X as an ordered sequence $X:D \rightarrow \{F,T\}^n$ of states. We will often deal with Boolean expressions built up from atomic parts of the form $(\exists^x t)\Delta$, or equivalently, from parts of form $(\forall^x t)\Delta$. In this connection the following notation is useful.

$\sup^x Z = \{a; (\exists^x t)Zt = a\}$ = smallest set S of states, $(\forall^x t)Zt\in S$.

Thus, $\sup^x Z$ consists of those states of Z which occur cofinally to x. The reader will establish the following lemma; a general version of it will be proved in Section 5.

<u>Convention</u>! Whenever x is used in either notation $(\forall^x t)$, $(\exists^x t)$, $\sup^x$, it will range over limit-ordinals only. Thus, in an expression such as $(\forall x)\mathcal{C}[\sup^x Z, Zx]$, $(\forall x)Zx \equiv (\exists^x t)Zt$ the reader will read $(\forall x)$ as "for all limits x". With von Neumann we will identify the ordinal α with the set of all ordinals $t<\alpha$. So $t<\alpha$ means $t\in\alpha$, and $\alpha' = \alpha\cup\{\alpha\}$.

Lemma 1.2: a) Every condition of either form $(\exists^x t)L[Zt]$, $(\forall^x t)L[Zt]$ can be stated in the form $\mathcal{U}[\sup^x Z]$.

b) A disjunction of conditions $\mathcal{U}_1[\sup^x Z_1]$, $\mathcal{U}_2[\sup^x Z_2]$ can be stated in the form $\mathcal{U}[\sup^x < Z_1, Z_2 >]$.

c) Every condition $\mathcal{U}[\sup^x Z]$ can be restated as a Boolean expression in conditions of the form $(\exists^x t)L[Zt]$. Hence, $\sup^x$-conditions are exactly the Boolean combinations of $(\exists^x t)$-conditions (i.e., $(\forall^x t)$-conditions).

The following lemma contain's Skolem's idea of replacing bounded quantifiers by recursions.

Lemma 1.3: On any ordinal α, $Zy \equiv (\forall t)^y E(t)$ defines the same predicate Z as does the recursion $Z0 \equiv T$, $Zt' \equiv [Zt \wedge E(t)]$, $Zx \equiv (\forall^x t)Zt$. Similarly one can replace $Zy \equiv (\exists t)^y E(t)$ by the recursion $Z0 \equiv F$, $Zt' \equiv [Zt \vee E(t)]$, $Zx \equiv (\exists^x t)Zt$.

The proof is left to the reader. We are now ready to state prenex-form lemmas for ordered and well-ordered systems.

Lemma 1.4 (prenex form for linear order): Every MT-formula Σ without free individual variables in the only primitive $<$ is equivalent, on the base of the axioms for linear order, to a formula of the form

$$(\text{predicate prefix}).(\forall x_1)(\forall x_2)^{x_1}\dots(\forall x_n)^{x_{n-1}}M \wedge (\exists x_1)(\exists x_2)^{x_1}\dots(\exists x_m)^{x_{m-1}}Q.$$

Here M and Q are matrices in which $<$ does not occur, and Q

contains predicate variables from the prefix only, and contains no individual variables except $x_1,\dots,x_m$.

Proof: Using lemma 1.1 we may assume that Σ is of the form (predicate prefix).$(\forall x_1 \dots x_n) M(x_1,\dots,x_n) \wedge (\exists x_1 \dots x_m) Q(x_1,\dots,x_m)$. Because of the axioms of linear order we can replace the part $(\forall x_1 \dots x_n) M(x_1,\dots,x_n)$ by $(\forall x_1)(\forall x_2)^{x_1} \dots (\forall x_n)^{x_{n-1}} \overline{M}(x_1,\dots,x_n)$, whereby $\overline{M}(x_1,\dots,x_n)$ is the conjunction, over all $\{i_1,\dots,i_n\} \subseteq \{1,\dots,n\}$, of the terms $M(x_{i_1},\dots,x_{i_n})$. And now we may replace $\overline{M}(x_1,\dots,x_n)$ by $\overline{\overline{M}}(x_1,\dots,x_n)$, which results from $\overline{M}$ by replacing each occurrence of $x_i<x_j$ by T, if $n \geq i > j$, and by F if $i \leq j \leq n$. Note that in $\overline{\overline{M}}$ the letter $<$ has no occurrence. The dual procedure will put $(\exists x_1 \dots x_m) Q$ into the required form. Q.E.D.

Lemma 1.5 (prenex form for well-orders): Let $\Sigma(X)$ be an MT-formula in the only free variables $X = \langle X_1,\dots,X_n\rangle$, and the only primitives $<$, 0, $'$. Let α be any ordinal. In the system $[\alpha, 0, ', <]$, Σ is equivalent to a formula of form (predicate prefix in Y)$(\exists Z)$.$A[Zo] \wedge (\forall t) B[Xt, Yt, Zt, Zt'] \wedge$ $\wedge (\forall x) \mathcal{C}[\sup^x Z, Zx] \wedge D[Z\alpha]$. Here Y is a sequence of predicate variables $Y_1,\dots,Y_s$, each occurring in the prefix. Z is a sequence $Z_1,\dots,Z_m$ of predicate variables, ranging over subsets of α'; so that $Z\alpha$ is defined. A, B, $\mathcal{C}$, D are matrices in the indicated atomic parts. t ranges over α, x ranges over limits $\leq \alpha$. So, if α is a limit $C[\sup^{\alpha} Z, Z\alpha]$ is part of the condition expressed by the formula, and if $\alpha = \beta'$, $B[X\beta, Y\beta, Z\alpha]$ is part of this condition. (Thus, the prenex form is not quite an MT-formula. By considering separately the cases where α is, and is not a limit one can easily remedy this situation.)

Proof: We may assume that 0 and ' do not occur in $\Sigma(X)$, else these primitives can be replaced by their definitions in well-ordered systems. Next we use Lemma 1.4 to put $\Sigma(X)$ into prenex form. Assume for example $n = 3$ and $m = 2$, so that $\Sigma(X)$ becomes

$$(\text{prefix in } Y).\ (\forall x_1)^{\alpha}(\forall x_2)^{x_1}(\forall x_3)^{x_2} M(X, Y, x_1, x_2, x_3) \wedge (\exists x_1)^{\alpha}(\exists x_2)^{x_1} Q(Y, x_1, x_2).$$

Next we put M into conjunctive form $\bigwedge_k M_k$, whereby each M_k is of form

$$M_{k1}[Xx_1, Yx_1] \vee M_{k2}[Xx_2, Yx_2] \vee M_{k3}[Xx_3, Yx_3] .$$

We process Q in the dual form, and can now move the individual quantifiers inside. $\Sigma(X)$ is equivalent to,

$$(\text{prefix } Y).\ \bigwedge_k (\forall x_1)^{\alpha}[M_{k1}(x_1) \vee (\forall x_2)^{x_1}[M_{k2}(x_2) \vee (\forall x_3)^{x_2} M_{k3}(x_3)]] \wedge$$

$$\bigvee_h (\exists x_1)^{\alpha}[Q_{h1}(x_1) \wedge (\exists x_2)^{x_1} Q_{h2}(x_2)] .$$

We now introduce new predicate variables Z_{k1}, Z_{k2}, Z_{k3} for each k and Z_{h1}, Z_{h2} for each h (assume that all k's are different from all h). Call the following formulas the "definitions" of the Z's:

$$Z_{k3}x \equiv (\forall t)^x M_{k3}(t) \qquad\qquad Z_{h2}x \equiv (\exists t)^x Q_{h2}(t)$$

$$(1) \qquad Z_{k2}x \equiv (\forall t)^x [M_{k2}(t) \vee Z_{k3}t] \qquad\qquad Z_{h1}x \equiv (\exists t)^x [Q_{h1}(t) \wedge Z_{h2}t]$$

$$Z_{k1}x \equiv (\forall t)^x [M_{k1}(t) \vee Z_{k2}t]$$

Our formula now is clearly equivalent to

$$(\text{prefix } Y)(\exists Z).(\forall x)^{\alpha+1}[\text{definition of } Zx] \wedge [\bigwedge_k Z_{k1}\alpha \wedge \bigvee_h Z_{h1}\alpha]$$

The second conjunction is the expression $D[Z\alpha]$ we are looking for. To obtain the other matrices $A[Z0]$, $B[Xt,Yt,Zt,Zt']$, $\mathcal{C}[\sup^x Z, Zx]$ we first use Lemma 1.3, to replace the definitions (1) by equivalent recursions. $A[Z0]$ is the conjunction, $Z0 = a$, of the initial conditions. $B[Xt,Yt,Zt,Zt']$ is the conjunction, $Zt' = F[Xt,Yt,Zt]$ of the transition conditions from t to t', and for $\mathcal{C}$ we have the conjunction of the transition conditions at limits x. This $\mathcal{C}$ is therefore a Boolean expression in parts of the form $(\forall^x t)Z_i t$, $(\exists^x t)Z_i t$, Zx. By Lemma 1.2 it follows that $\mathcal{C}$ is also of the required form $Zx = \mathcal{B}[\sup^x Z]$.

Q.E.D.

It should be noted that all these forms of the prenex-form lemma hold also in WMT. In the proof of Lemma 1.5 we have actually seen that Z is deterministic, in the sense that it satisfies a recursion $Z0 = a$, $Zt' = F[Xt, Yt, Zt]$, $Zx = \mathcal{G}[\sup^x Z]$. Thus $\Sigma(X)$ takes the form

(prefix Y)[the $Z = \mathrm{Rec}(X,Y)$ satisfies the output-condition $D[Z\alpha]$].

This very natural relation of finite-state-recursions and MT's was first observed, by the author, in the case of WMT$[\omega, 0, ']$, and led to the various decision methods for MT's, using finite automata. We will outline the basic method in the next section.

We terminate this section with some explanatory remarks concerning the monadic theory of linear orders and well-orders. Clearly isomorphic structures $\underline{D}_1 \cong \underline{D}_2$ are MT-equivalent (in fact L-equivalent for arbitrarily strong languages L). The most famous example, where the implication also goes the other way around, is the following. The Peano-axioms Pea_ω are categorical, and are MT-sentences. Hence, $[\omega, 0, ']$ is MT-categorical and finitely MT-axiomatizable. This classical situation generalizes to ordinals $\alpha < \omega^\omega$, and to ordinals ω_n, but things are quite different at ω^ω:

a) For every ordinal $\alpha < \omega^\omega$, $[\alpha, <]$ is MT-categorical and finitely MT-axiomatizable.

b) For every ordinal $\omega^\omega \leq \alpha < \omega_1$, $[\alpha,<]$ is neither MT-categorical nor finitely MT-axiomatizable. In fact, for each $\beta < \omega^\omega$ and each $1 \leq \gamma < \omega_1$, $[\omega^\omega\gamma+\beta,<]$ is MT-equivalent to $[\omega^\omega+\beta,<]$.

c) For every n, $[\omega_n,<]$ is MT-categorical and finitely MT-axiomatizable (if AC).

Of these statements, clearly b) is the hard one to prove; we need the results in Section 4. For example, the fact that all theories $MT[\alpha,<]$, $\alpha<\omega_1$ are decidable implies that only countably many of them can be different (there are but countably many decision methods). So, many of the $[\alpha,<]$, $\alpha<\omega_1$ must be MT-equivalent. To show a) one easily makes up finite categorical axiom systems Pea_α, for each $\alpha<\omega^\omega$. In the case of c) the following formulas are used to make up the necessary finite axiom systems Pea_{ω_n}. The key idea is that in MT{linear orders} one can make accessibility statements, and due to the axiom of choice, these become cardinality statements.

Let Σ be an MT-sentence in the only primitive letter <. The relativization of Σ to x is the formula $\Sigma[x]$ obtained as follows:

1. Replace all occurrences of x in Σ by the first individual variable which does not occur in Σ.

2. Replace all individual quantifiers $(\forall t)$, $(\exists t)$ by $(\forall t)^x$, $(\exists t)^x$.

3. Replace all set-quantifiers $(\forall U)\Delta$ by $(\forall U)[(\forall t)[Ut \supset t<x] \supset \Delta]$ and $(\exists U)\Delta$ by $(\exists U)[(\forall t)[Ut \supset t<x] \wedge \Delta]$.

Similarly one constructs the relativized $\Sigma[X]$ of Σ to X:

1. Replace all occurrences of X in Σ.

2. Replace individual quantifiers $(\forall t)\Delta$, $(\exists t)\Delta$ by $(\forall t)[Xt \supset \Delta]$, $(\exists t)[Xt \wedge \Delta]$.

3. Replace set quantifiers $(\forall U)\Delta$, $(\exists U)\Delta$ by $(\forall U)[U \subseteq X \supset \Delta]$, $(\exists U)[U \subseteq X \wedge \Delta]$.

We now make up a succession of MT-sentences, all in the only primitive letter $<$. The meaning of these sentences, when interpreted in the structure $[\alpha,<]$, α any ordinal, is indicated at the right.

$x=y$	:	$(\forall U).Ux \supset Uy$	Leibnitz' definition of equality
Wlo	:	conjunction of the axioms of well-order	
Zer	:	$\sim(\exists t)t=t$	α is zero
Suc	:	$(\exists x)\sim(\exists t)x<t$	α is a successor
Cof(U)	:	$(\forall x)(\exists t)[Ut \wedge x<t]$	U is cofinal subset of α
$Acc_{<0}$	:	Zer $\vee$ Suc	α is of accessibility 0 or 1
Kap_n	:	$Wlo \wedge (\forall x)Acc_{<n}[x] \wedge \sim Acc_{<n}$	α is ω_n
Acc_n	:	$(\exists U)[Cof(U) \wedge Kap_n[U]]$	α is ω_n-accessible
$Acc_{<n+1}$	:	$Acc_{<n} \vee Acc_n$	α is of accessibility $<\omega_{n+1}$

Note that $Kap_0 \succ \mathfrak{D}$ says "$\mathfrak{D} = [D,<]$ is a well-order in which every element x is first or a successor, and $\mathfrak{D}$ is neither empty nor has a last member". Thus, Kap_0 is a categorical MT-axiom for $[\omega_0,<]$ (an alternative to Peano's axioms Pen_{ω_0}). Therefore, $Acc_0 \succ \mathfrak{D}$ says "$\mathfrak{D}$ contains a cofinal ω_0-sequence". In particular, Acc_0 holds in $[\alpha,<]$ just in case α is an ω_0-limit, which we also express by "α <u>is</u> ω_0-<u>accessible</u>", or "<u>the</u> <u>accessibility</u> <u>of</u> α <u>is</u> ω_0". The accessibility of a successor ordinal we define to be 1, and the accessibility of 0 is 0. So, $Acc_{<1} \succ [\alpha,<]$ just says "α is of either accessibility 0, 1, or ω_0. Hence, $Kap_1 \succ \mathfrak{D}$ means "$\mathfrak{D}$ is (isomorphic to) the first ordinal $\varkappa_1$ which is neither of accessibility 0, 1, nor ω_0". In other words Kap_1 is a categorical MT-axiom for $[\varkappa_1,<]$. Similarly, Kap_2 categorically describes the first ordinal $\varkappa_2$ which is not 0, 1, ω_0, or $\varkappa_1$-accessible, etc.

It is easy to show that $\omega_n \leqq \varkappa_n$, and a familiar argument, using AC^{γ}_{γ}, where $\gamma = \omega_{n-1}$, shows that ω_n is $\varkappa_n$-accessible. Hence, if AC then $\omega_n = \varkappa_n$, and Kap_n becomes the categorical axiom system for ω_n. Note that ω_ω is of accessibility $\varkappa_0 = \omega_0$, and that, with AC, $\varkappa_\omega = \omega_{\omega+1}$.

2. A method for eliminating set-quantifiers

The proofs of decidability presented in this paper afford further examples of what seems to be a quite general method for the elimination of quantifiers in monadic second order theories. In this section we will describe the method, outline its use in various cases, and state unsolved problems which seem accessible by it.

Today it is clear to every student of logic, and even to many mathematicians, that the nested occurrence of quantifiers has much to do with the intricacies of mathematical statements. If this nesting is not too complex, we may be able to understand (intuitively grasp) the meaning of a statement. One is tempted to claim: The only way to understand a statement Σ (or a notion $\Sigma(X)$) is by elimination of quantifiers; that is, by finding related statements Σ' in which the nesting is simple. Much of Skolem's work consists of contributions to this art, and some of it he learned from Löwenheim. In particular, there is the Skolem-Löwenheim method of finding decision methods for particular theories. One of the first theories subjected to this method was MT[K], K the class of all structures consisting of just a domain D [16, 18]. Our method, in its general form, simply is the following version of quantifier elimination. We state matters, in case of a theory $T = MT[\underline{D}]$.

Item 1: Find a class of formulas, called *the kernels*, such that every formula $\Sigma(X)$ of the theory T under investigation can be put into prenex form (prefix $Y)\Gamma(X,Y)$, whereby Γ is a kernel.

Item 2: Try to establish *the complementation-lemma* for the set of kernels: To every kernel $\Gamma(X,Y)$ one can find another kernel $\Gamma'(X,Z)$ such that $\sim(\exists Y)\Gamma(X,Y)$ is equivalent to $(\exists Z)\Gamma'(X,Z)$.

Item 3: *The decision problem* for kernels: Try to find a method which applies to every kernel $\Gamma(Z)$, and which tells whether or not $(\exists Z)\Gamma(Z)$.

Here X, Y, Z are tuples of variables (at this stage not necessarily set-variables). The art of the game of course is to choose kernels $\Gamma(X)$ so that their meaning is intuitively clear. Note that there is much freedom here, as the kernels need not be formulas of the theory T. Suppose items 1 and 2 have been established for T. Then we have,

The definability theorem: Every relation R on $\mathcal{P}D$ definable by a formula $\Sigma(X)$ in the theory T is of form $\hat{X}(\exists Z)\Gamma(X,Z)$, whereby Γ is a kernel. In particular, every sentence Σ of the theory T can be put into the form $(\exists Z)\Gamma(Z)$, whereby Γ is a kernel.

Proof: By item 1, $\Sigma(X)$ can be put in prenex form, say for example $(\exists Y_1)(\forall Y_2)(\exists Y_3)\Gamma_1(X,Y_1,Y_2,Y_3)$, ($X,Y_1,Y_2,Y_3$ tuples of variables), Γ_1 a kernel. By the complementation-lemma (item 2), $\sim(\exists Y_3)\Gamma_1(X,Y_1,Y_2,Y_3)$ is of form $(\exists Z_3)\Gamma_2(X,Y_1,Y_2,Z_3)$, Γ_2 a new kernel. Hence $\Sigma(X)$ now is expressed in the form $(\exists Y_1)\sim(\exists Y_2 Z_3)\Gamma_2(X,Y_1,Y_2,Z_3)$. Using the complementation-lemma we find a kernel Γ_3 such that $\sim(\exists Y_2 Z_3)\Gamma_2(X,Y_1,Y_2,Z_3)$ is equivalent to $(\exists Z_2)\Gamma_3(X,Y_1,Z_2)$. Hence, $\Sigma(X)$ is put into form $(\exists Y_1 Z_2)\Gamma_3(X,Y_1,Z_2)$.

Q.E.D.

Definability problems about T are thus reduced, by items 1 and 2, to existential questions about kernels. If also item 3 can be handled we have a decision method for truth of sentences in T.

Although these ideas are not limited to monadic second order theories, they have been particularly useful in this case. For MT[D] we have available the prenex form Lemma 1.1. Depending on D this can be further improved, see for example Lemmas 1.4 and 1.5. At this point very natural kernels suggest themselves, in the form of transition systems or finite-state-recursions. This reduction of problems about MT's to problems about appropriately chosen finite automata is an integral part of our method.

Let us think of individual variables as ranging over time. A k-tuple X of unary predicates becomes a time-sequence Xt of states in the finite set $\{F,T\}^k$. A kernel $\Gamma(X,Y)$ is called a finite transition-system (or non-deterministic automaton); with input states the states of X. The states of Γ are the states of Y. $\Gamma(X,Y)$ is read as "the sequence Y of states of Γ is a run of Γ on the sequence X of input-states of Γ", or shortly "Y is an X-run of Γ". If there is an X-run of Γ, i.e., if $(\exists Y)\Gamma(X,Y)$ we will say "Γ accepts the input-sequence X". The set consisting of all accepted input-sequences will be called the behavior of the transition-system Γ. In this terminology the complementation lemma says: The complement of the behavior of a finite transition-system still is the behavior of a finite transition-system. It yields: Every formula $\Sigma(X)$ represents the behavior of a finite transition-system. In particular, a sentence Σ represents the behavior of an input-free system $\Gamma(Y)$. That is: Σ is true just in case the input-free system Γ admits a run (i.e., $(\exists Y)\Gamma(Y)$).

Our method for investigating the monadic theory MT[D] of a structure D therefore consists of the following:

Item 1: Put the kernels $\Gamma(X,Y)$ into a form which makes concrete the terminology of the run of transition-systems.

Item 2: (the complementation lemma) Show that the complement of the behavior of a transition-system still is the behavior of a transition-system. (Questions about definability in the theory now are reduced to questions about behavior of transition-systems.)

Item 3: Find a method for deciding whether or not an input-free transition-system $\Gamma(Y)$ admits a run. (This yields a decision method for truth of MT-sentences in D.)

An important variant of this method, the deterministic method, uses finite automata (deterministic transition-systems). One version of this may be described as a method for attacking the complementation lemma (item 2). Namely, for various structures D one can find a natural class $Z = A(X)$ of recursive definitions of state-sequences Z from input-sequences X, called finite-state recursions. The objective now is to add a simple class of "output conditions" $\Omega(Z)$, which must be obviously closed under complementation. An input-sequence X is accepted by the automaton $[A,\Omega]$ just in case $Z = A(X)$ satisfies the output-condition $\Omega(Z)$, i.e., in case $(\exists Z)[Z = A(X) \wedge \Omega(Z)]$. The behavior of $[A,\Omega]$ is the set of accepted X. In this deterministic case the complementation lemma becomes trivial, as the behavior of $[A,\sim\Omega]$ is the complement of the behavior of $[A,\Omega]$. To prove item 2, it therefore suffices to show:

Item 2': (the subset construction) To every transition-system $\Gamma(X,Y)$ one can construct an automaton $[A,\Omega]$ of equal behavior.

Of course one also must take care that $Z = A(X) \wedge \Omega(X)$ is (or is obviously equivalent to) a kernel.

Another version of the deterministic method is to operate with deterministic kernels, only. The important fact to be proved in this case is the projection lemma.

Item 1": Show that every formula $\Sigma(X)$ can be put into the form (prefix Y)$(\exists Z).Z = \mathcal{A}(X,Y) \wedge \Omega(Z)$, for an appropriately chosen class of recursions, $\mathcal{A}$ (called the automata), and a class of output-conditions Ω which is closed under complementation.

Item 2": (the projection lemma) Show that the projection $(\exists Y)B(X,Y)$ of the behavior $B(X,Y)$ of an automaton $[\mathcal{A}(X,Y),\Omega]$ still is the behavior $B'(X)$ of an automaton $[\mathcal{A}'(X),\Omega']$.

Item 3": Find a method for deciding whether or not the run Z of an input-free automaton $[\mathcal{A},\Omega]$ satisfies the output-condition Ω.

Because the behaviors of deterministic systems are obviously closed under complementation, the projection lemma implies its dual: If $B(X,Y)$ is the behavior of an automaton then so is $(\forall Y)B(X,Y)$. Hence items 1" and 2" yield a method for eliminating quantifiers in a prenex form such as

$$(\exists Y_1)(\forall Y_2)(\exists Y_3)(\exists Z)[Z = \mathcal{A}(X,Y_1,Y_2,Y_3) \wedge \Omega(Z)] .$$

The result is: Every formula $\Sigma(X)$ (of the monadic theory under investigation) is equivalent to the behavior $B(X)$ of an automaton $[\mathcal{A},\Omega]$. In particular sentences Σ correspond to input-free automata $[\mathbf{A},\Omega]$, in which the recursion $\mathbf{A}$ defines a time-sequence Z of states, outright. Σ is true (in the theory under investigation just in case the unique run Z of $\mathbf{A}$ satisfies $\Omega(Z)$. Hence, item 2" yields a decision method for the theory.

We will now survey the monadic theories which are known decidable as of today (Sept. 1971). All these results can be proved using our method. Excepting Ehrenfeucht and Läuchli, all these results have actually been obtained using our method, although none of the authors acknowledges this fact.

1. WMT[ω,<]: In his 1957-58 lectures at Illinois, A. Church used a simple sort of monadic predicate recursion $Z = \mathbf{A}(X)$ to represent (ordinary) finite automata. He also used simple sorts of first-order formulas $\Sigma(X,Y)$ to represent input-to-output conditions on such automata. Attending these lectures, it occurred to us that MT[ω,<] (equivalently MT[ω,0,']) should be considered as a more natural condition-language, and we designed the deterministic method for eliminating quantifiers. At this time a subset-construction of Myhill, and a projection lemma of Medvedev, both in the sense of Kleene's behaviors of automata, were available. However, these seemed more likely to serve as item 2', and item 2" for WMT[ω,<], rather than the full monadic theory. We communicated these ideas to C.C. Elgot and others. The actual proofs of decidability, using the deterministic method, were obtained independently; see [1] and [10]. In [11] this result is credited to A. Ehrenfeucht, and general

methods are provided for stepping up such decision methods. (These methods should be considered in the case $MT[\omega_1,<]$, and in Rabin's case [14].) The problem of Tarski's, to show $MT[\omega,0,']$ decidable, remained unanswered.

2. $MT[\omega,<]$: Up to choice of primitives this is $MT[\omega,0,']$, and is just right for stating Peano's axioms. In his work on incompleteness Gödel used monadic functions in place of monadic predicates. We call this theory $SMT[\omega,<]$ (strong monadic theory); using Gödel's work, Church showed this theory recursively undecidable. In [2] we showed $MT[\omega,<]$ is decidable; as we were unable to extend the subset construction to ω-sequences, we designed and used the non-deterministic method.

Transition-systems of form

$$\Gamma(X,Z): \quad E[Z0] \wedge (\forall t)H[Xt,Zt,Zt'] \wedge \Omega(Z),$$

are used to serve as item 1. As output-condition Ω, formulas of form $(\exists^{\omega}t)L[Zt]$ are used. E,H,L are propositional expressions. Thus E and L should be interpreted as sets of states of Z, and H means a relation between the present input-state Xt, the present state Zt, and the next state Zt' of the system. Thus, Z is an X-run through Γ, in case its initial state Z0 belongs to E, at each instance t the transition-relation H holds, and Z runs through L infinitely often. The reader should visualize this situation as suggested by the figure on the following page.

Bü 2

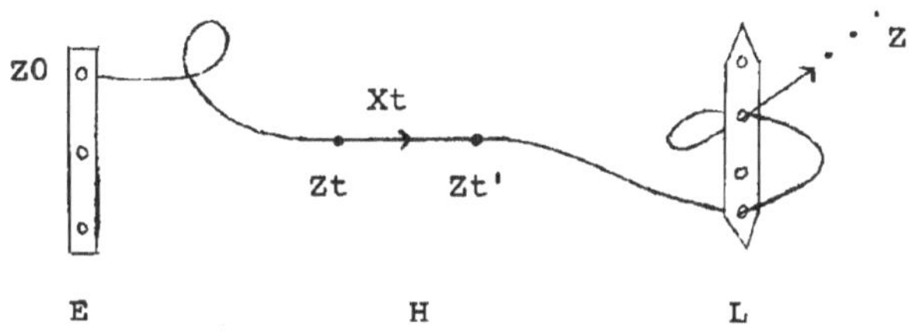

The complementation lemma (item 2) was proved by using a combinatorial theorem of Ramsey's. The decision method (item 3) is a simple matter. Shoenfield (unpublished) found a proof of the complementation lemma which uses König's infinity lemma.

A very important contribution to the field is contained in [13]. McNaughton extends the subset-construction (or the projection lemma) to ω-behavior of finite automata $Z0 = s_0$, $Zt' = F[Xt, Zt]$, and thus makes available the deterministic method for elimination of quantifiers. The output-conditions $\Omega(Z)$ now must be closed under complementation. Therefore it is natural to work with Boolean combinations of expressions of form $(\exists^{\omega} t)L[Zt]$. It is easy to see that such conditions $\Omega(Z)$ can be restated in the more pleasant form $\sup Z \in \mathfrak{F}$, whereby $\mathfrak{F}$ is a set of sets of states of Z, and $\sup Z$ stands for the set of those states which Z takes on infinitely often.

McNaughton's construction is quite intricate. In Section 3 we will present a more explicit version, which will generalize naturally to all countable ordinals (Section 4). We will also present a new version of our original proof of the complementation lemma. It does not use either Ramsey or König, and therefore yields ideas which will be useful at ω_1 (Section 6).

3. $WMT[\alpha,<]$ for any ordinal α: In [4] the deterministic method is extended from the case $\alpha = \omega$, to work for any ordinal α. This also yields decidability of WMT{all well-orders}, and every $\alpha \geq \omega^{\omega} \cdot 2$ is WM-equivalent to some $\omega^{\omega} \leq \beta < \omega^{\omega} \cdot 2$. The trick was to extend the special finite-state recursions in [1] to transfinite ordinals $x \leq \alpha$, and to use propositional formulas $U[Z\alpha]$ as output-conditions. The point here is that complexities, which otherwise would enter into the form of the output-conditions are automatically handled by the last transition from $\{t; t<\alpha\}$ to α. The result is that the same form of the projection lemma (and of the subset construction, and hence the complementation lemma) work uniformly for all α. This uniformity is essential, as the proof (that the construction works) goes by induction on α. We do not know a non-deterministic proof in this case, because it is not clear how to obtain a uniform complementation lemma, without using the deterministic method.

4. $MT[\alpha,<]$ for countable ordinals α: In Section 4 we show that each of these theories is decidable, and so is MT{all countable well-orders}, and for countable ordinals $\alpha > \omega^{\omega} \cdot 2$ there is a $\omega^{\omega} \leq \beta < \omega^{\omega} \cdot 2$ such that α and β are MT-equivalent.

Item 1: It is natural to use as transition systems (kernels) expressions of form

$$E[Z0] \wedge (\forall t)^{\alpha} H[Xt, Zt, Zt'] \wedge (\forall x)^{\alpha'} \mathcal{K}[\sup^{x} Z, Zx] \wedge \mathcal{U}[Z\alpha].$$

Here x ranges over all limits $\leq\alpha$, and $\sup^x Z$ stands for the set of all states which Z takes on cofinal x. Note that Z is extended to α, and the output condition is simply a set of states of Z (a propositional connective), an idea which is taken over from <u>3</u>.

<u>Item</u> 2': The complementation lemma is proved by using the deterministic method. The automata recursions take the form $Z0 = s_0$, $Zt' = F[Xt,Zt]$, $Zx = \mathcal{S}[\sup^x Z]$, and are extended through to α. The terminal conditions again take the simple form $U[Z\alpha]$. The subset construction is a natural extension of that presented for $\alpha=\omega$, in Section 3. In fact, the proof at ω becomes the prototype for the inductive step at limits α, in the extended subset construction.

<u>Item</u> 3': It is not difficult, for any α, to find a method for deciding whether or not $U[Z\alpha]$, if Z is given by an input-free recursion $Z0 = s_0$, $Zt' = F[Zt]$, $Zx = \mathcal{S}[\sup^x Z]$.

As mentioned in <u>3</u>., it is important that the subset-construction does not depend on α, and we therefore get one form of the complementation lemma, which applies uniformly to all countable α. In contrast, the non-deterministic method used for $\alpha=\omega$ in Section 3, will yield a complementation lemma at α only if we already possess a uniform way of handling all $t<\alpha$ (and this uniformity will be disrupted at α). So here is one more situation where the subset-construction is essential. For others see [5, 6].

5. $MT[\omega_1,<]$: In the weak monadic case ω_1 is indistinguishable from ω^ω, and so is ω_β for any $\beta\geq 1$. In the monadic case things become more interesting. For one, we now can make accessibility statements. It is easy to make up MT-sentences A_n such that A_n holds in $[\alpha,<]$, if and only if, $\omega_n \Rightarrow \alpha$ (α is ω_n-limit). Let $A_n[x]$ be the formula obtained from A_n by relativizing all individual and set-quantifiers to x. Now, the formula $A_n[x] \wedge (\forall t)^x {\sim} A_n[t]$ defines the ordinal ω_n, in the monadic theory of all ordinals, and the sentence $A_n \wedge (\forall x){\sim}A_n[x]$ holds in $[\omega_n,<]$ but is false in every other ordinal $[\alpha,<]$. Hence $MT[\omega_n,<]$ is categorical, in the sense that there is just one standard model. Besides these accessibility statements there are monadic statements about linear orders, which can best be expressed via filters of cofinal closed sets (see Section 5), and which are much more sophisticated. These filters seem to play an essential role in the formulation of transition-conditions from all $t<x$ to the limit x. Which of the filters comes to use depends on the accessibility of x.

The filter $\mathscr{I}_0^x$ of all end segments of x was already used in the transition to countable limits x (which are ω_0-accessible). Namely $\sup^x Z$ is just the smallest set D of states of Z, such that $\{t; t<x \wedge D[Zt]\} \in \mathscr{I}_0^x$. It turns out that $\sup_0 = \sup$ cannot be used to serve in the transition conditions to ω_1, or in the output condition at ω_1. We therefore

introduce the filter $\mathcal{J}_1$ of all cofinal closed subsets of ω_1, and we define $\sup_1 Z$ to be the smallest set D of states of Z, such that $\{t; t<\omega_1 \wedge D[Zt]\} \in \mathcal{J}_1$. This notion is just right to serve in output conditions at ω_1, for transition systems. It is still too weak in the deterministic case (in fact, in this case a $\sup_1$-condition may be replaced by a $\sup_0$-condition, by enlarging the set of states of Z). We therefore will handle the decision problem for ω_1, in the non-deterministic manner. Our method at ω_1 is modeled after the non-deterministic method at ω_0 (see Section 3), and is outlined thus.

Item 1: As transition systems we use

$$E[Z0] \wedge (\forall t)H[Xt,Zt,Zt'] \wedge (\forall x)\mathcal{K}[\sup_0^x Z, Zx] \wedge \mathcal{U}[\sup_1 Z].$$

Item 2: The complementation lemma is proved by using the results about ordinals $x<\omega_1$ (Section 4), and the fact that $\mathcal{P}\omega_1/\mathcal{J}_1$ has no atoms (Section 5).

Item 3: It is not difficult to decide whether or not a run Z exists in an input-free system

$$E[Z0] \wedge (\forall t)H[Zt,Zt'] \wedge (\forall x)\mathcal{K}[\sup_0^x Z, Zx] \wedge \mathcal{U}[\sup_1 Z] \quad .$$

6. As, in Section 6, we handle the transition to ω_1 in the non-deterministic manner, this proof will not take us far past ω_1.

The trouble is the matter of uniformity, which we have already mentioned in 3. and 4. However, we now know how to extend the deterministic method to ω_1-accessible limits, and we therefore have a uniform way of handling ordinals $\alpha<\omega_2$. Consequently, $MT[\alpha,<]$ is decidable for all $\alpha<\omega_2$. We will present the details of this proof in another paper. What is needed is more work on cofinal closed sets; in particular on the relation between the cofinal closed subsets of $[\omega_1,<]$ and those of $[S,<]$, S a cofinal subset of ω_1. If $\omega_1 \Rightarrow x$ and S is cofinal x, let $\sup_1^S Z$ denote the smallest set D of states of Z such that $\{t; t\in S \wedge D[Zt]\}$ contains a cofinal closed subset of $[S,<]$. If $L[Zt]$ is a matrix let $\sup_1^{L,x}Z$ be $\sup_1^S Z$, whereby $S = \{t; t<x \wedge L[Zt]\}$. The automata to be used for $\alpha<\omega_2$ then take the form

$$
(1)\qquad
\begin{aligned}
Z0 &= a,\\
Zt' &= H[Xt,Zt],\\
Zx &= \mathcal{J}_0[\sup_0^x Z], \quad \text{if } \omega_0 \Rightarrow x\\
Zx &= \mathcal{J}\,[\sup_1^x Z, \sup_1^{L_1,x} Z, \ldots, \sup_1^{L_h,x} Z], \quad \text{if } \omega_1 \Rightarrow x.
\end{aligned}
$$

Here $L_1[Zt],\ldots,L_h[Zt]$ are matrices.

In fact we do not see any new difficulties at ω_{n+1}-accessible limits. It is clear that the filter $\mathcal{J}_{n+1}$ of cofinal sets which are closed under ω_n-limits will come to use. However, due to the necessity of relativization to $L[Zt]$'s, things become rather involved, technically. We are prepared to conjecture that $MT[\alpha,<]$ can be shown decidable for any α smaller than the first limit at which one of our filters $\mathcal{J}_n$ admits an atom.

7. $MT[\eta,<]$, η the order of rationals: Let T_η stand for this theory and let $T = MT\{$all countable linear orders$\}$. The decision problems for these theories are equivalent. Namely, let A be the axiom for dense linear orders. Then, for any MT-sentence Σ, $\Sigma \in T_\eta$ if and only if $[A \supset \Sigma] \in T$; hence the decision problem for T_η reduces to that for T. The other way around we argue as follows. According to Cantor, every countable linear order is isomorphic to a subset of η. Hence, if $\Sigma[X]$ is the relativization to the set X of the MT-sentence Σ, we have $\Sigma \in T$ if and only if $(\forall X)\Sigma[X] \in T_\eta$. This reduces the decision problem for T to that for T_η. It should be noted that from a method for T one obtains decision methods for $MT\{$all countable well-orders$\}$, and of $MT[\alpha,<]$ for some (the definable ones, but not all) countable orders. It is also clear that in T_η one has available a formula $Cnv(X)$, which states "X is a convergent subset of η" (or X, $\sim X$ is a cut). Hence, T_η is a significant fragment of analysis (in [14] this matter is pursued in detail).

In [12] Läuchli presents a decision method for $WMT\{$all countable ordinals$\}$ (because the Skolem-Löwenheim theorem holds for WMT, one can here omit "countable"). His method is based on work of Fraïssé and Ehrenfeucht. One wonders whether the same ideas could be used to handle T or T_η. It seems clear that our deterministic method is not directly applicable, as recursions are not available on η. Whether the non-deterministic method can be put to work directly on T_η should be investigated, as it might lead to ideas about MT[λ,<], λ the ordering on all reals. More naturally adapted to our automata-method is the following indirect approach.

It often happens that the original primitives of a theory T are not the right ones for eliminating quantifiers. Moreover, elimination of quantifiers may well be more natural in a theory which is stronger in expressive power. In fact, divising the right primitives plays a very important role in the art of quantifier elimination. Now, everyone knows that analysis is the theory of the two-splitting infinite tree. This tree may be defined to be the free algebra $\mathcal{T}$ = [F,e,f,g] with one generator e (the root of the tree) and two unary functions f, g (fx the right-successor, gx the left-successor of the vertex $x \in F$). Furthermore, the natural definition Ord(f,g,x,y) of an η-order $x<y$ on F happens to be an MT-formula (and it is equally simple to obtain a WMT-definition of < in $\mathcal{T}$). Therefore we suggested the problems to find decision methods for WMT[$\mathcal{T}$] and MT[$\mathcal{T}$]. The point of course was, that our method (then available in the one-successor case) seemed to make these problems accessible. Particularly so, because in $\mathcal{T}$ (as opposed to η) it is clear which sort of structures might be used to play the part of transition systems.

In 1965 J. Doner showed WMT[$\mathcal{T}$] decidable. This work is presented in [8]. Finite subsets of $\mathcal{T}$ are naturally interpreted as finite trees with labeled vertices, and it is obvious what a non-deterministic finite tree acceptor should be. However, as in the case $WMT[\omega,0,']$, the deterministic method is used. Doner's contribution was the idea of using automata (deterministic systems) which would scan finite trees starting at the frontier and ending at the root, rather than the other way around (the difference is that between the transition relations $Zt = F[Zft,Zgt,Xt]$ and $Zft = F_1[Zt,Xt]$, $Zgt = F_2[Zt,Xt]$). Once this clue, which is not obvious, is given, it is easy to generalize the proof from the one-successor case to that of two successors.

In [14] Rabin used our non-deterministic method to obtain a decision method for $MT[\mathcal{T}]$. While it is obvious that the transition relation of a system must be $H[Xt,Zt,Zft,Zgt]$, the problem of choosing the right terminal-conditions, among a host of reasonable ones, is staggering. Rabin's proof contains a series of combinatorial results, each comparable in intricacies to McNaughton's subset-construction. But then the result is a very powerful decision method, covering a significant fragment of analysis.

The existence of a powerful method for elimination of quantifiers is the best reason for considering monadic second order theory this author knows. Another feature is the flexibility available to naturally interpret other theories within a given MT. Witness our interpretation of $MT[\eta,<]$ in the MT of two successor functions and the many consequences of Rabin's result, stated in [14]. The following seem to be the most important cases, in which our method should be tried.

Problem 1: Find a decision method for $MT[\lambda,<]$, λ=ordering of the continuum.

More generally, for any ordinal α, let $2^{(\alpha)}$ consist of all well-ordered strings of two objects a, b and of length $<\alpha$. For such strings x, y let $x<_1y$ stand for "xa is an initial segment of y", and define $x<_2y$ correspondingly, for b in place of a.

Problem 2: Is $MT[2^{(\alpha)},<_1,<_2]$ decidable (a) for $\alpha=\omega+1$, (b) for every countable ordinal α?

Note that $\alpha=\omega$ is just the two-successor theory, handled by Rabin.

Problem 3: Show $MT[\Omega,<]$ is decidable; Ω the class of all ordinals.

Problem 4: How about (a) the MT of all linear orders, and (b) the MT of all unary functions?

This of course is related to Problem 2, and $MT[\eta_1,<]$, η_1 the ω_0-hyper-dense order (CH). Also, in case (a), our filters $\mathcal{J}_n^x$ of cofinal-closed sets will have to plan an essential role.

Addendum:

Recently, Shelah has announced a negative solution to problem 1 and therefore to problem 2. However it seems that the decision problem (item 3) for input free automata is still solvable for many (all?) countable α. It remains to be analyzed how much this means as to the decidability of problems on the real numbers.

Le Tourneau has shown that the MT of one arbitrary unary function is decidable, thus solving problem 4(b).

3. The working of the method in the case of ω_0

We will present here a streamlined proof of the complementation lemma for $MT[\omega_0,<]$, and an explicit version of McNaughton's subset-construction. Once these matters are clearly understood at ω_0, it will be easy to generalize to arbitrary countable ordinals.

Item 1: The kernels or transition-systems. We have already prepared things in Lemma 1.5. According to it, every formula $\Sigma(X)$ of our theory can be put into the prenex form

$$(\text{prefix } Y)(\exists Z).A[Z0] \wedge (\forall t)B[Xt,Yt,Zt,Zt'] \wedge \mathcal{C}[\sup Z].$$

Here and throughout this section $\omega=\omega_0$, and supZ stands for $\sup_{\omega_0} Z$, the set of all states a which Z takes on infinitely often. As kernels we will therefore choose those formulas which are of form,

(1) $\Gamma(X,Z) \quad : \quad E[Z0] \wedge (\forall t)H[Xt,Zt,Zt'] \wedge \mathcal{K}[\sup Z]$

Here X stands for a k-tuple, and Z for a n-tuple of set-variables, each ranging over all subsets of ω. E, H, $\mathcal{K}$ are matrices (i.e. propositional formulas) in the indicated parts. As desired we have,

<u>Lemma</u> 3.1: Every formula $\Sigma(X)$ of $MT[\omega,<]$ can be put into the form (prefix Y)$\Gamma(X,Y)$, whereby Γ is a kernel (or transition-system).

The second part of item 1, providing a concrete meaning for kernels, goes as follows.

The kernel (1) is a <u>finite transition-system</u> $\Gamma = [E,H,\mathcal{X}]$, consisting of a set of <u>input-states</u> $\{F,T\}^k$, a set of <u>states</u> $\{F,T\}^n$, an <u>initial set</u> E (subset of all states), a <u>transition-relation</u> H (relating one input-state with two states), and a <u>terminal set</u> $\mathcal{X}$ (set of sets of states). An <u>input</u> to Γ is an ω-sequence X=X0,X1,X2,... of input-states. An X-<u>run of</u> Γ is an ω-sequence Z of states such that $\Gamma(X,Z)$, i.e., Z0 belongs to the initial set E, the transition-condition H[Xt,Zt,Zt'] holds for all times $t\in\omega$, and Z satisfies the terminal condition $\sup Z\in\mathcal{X}$. In the figure we show the transition-system Γ as a labeled graph, whose vertices are the 2^n states of Γ. The vertices a and b are joined by a directed edge, labeled by the input state e, just in case H[e,a,b]. An X-run of Γ then is an ω-path Z through the graph, starting in E, "ending" with a sup in $\mathcal{X}$, and which carries in order the labels X0,X1,X2,... on its edges.

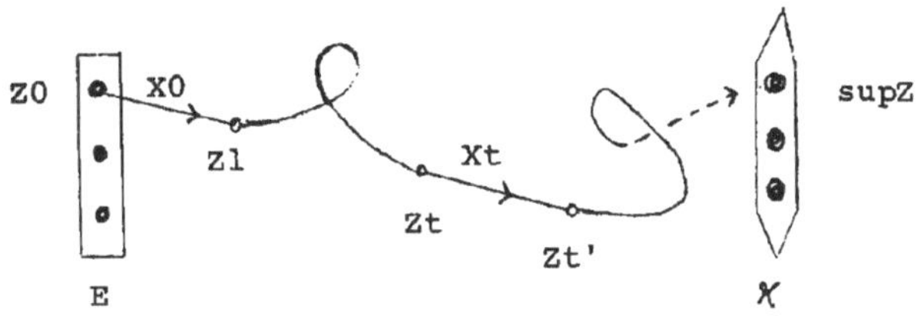

The input X __is accepted__ by the transition-system Γ just in case $(\exists Z)\Gamma(X,Z)$, that is, in case there is an X-run of Γ. We will also be using the following notations.

$X[u,v)$: the segment from u (inclusive) to v (exclusive) of a sequence X of states.

$Z[u,v]$: similarly, now inclusive Zv.

Depending on whether or not the full sequence X occurs in the context, $X[u,v)$ will mean the indicated part of X, or $X[u,v)$ will simply stand for a sequence of states of length $(v-u)$. Let $\Gamma = [E,H,\mathcal{K}]$ be a transition system, and let X be an input to Γ.

$Z[u,v]$ is an X-__run__ of H __from__ a __to__ b :

$$Zu=a \wedge (\forall t)_u^v H[Xt,Zt,Zt'] \wedge Zv=b .$$

In the theory of finite automata our transition graphs play an important role. We will not use any results proved in this theory. Rather, the methods and results discussed here may be considered a significant (if not the most significant) contribution to the theory of finite automata.

__Item__ 2: __The complementation lemma__. We will now prove a series of lemmas about the transition-system $\Gamma(X,Z)$ given by formula (1), leading to a second system $\Gamma'(X,S)$ which accepts exactly those

X not accepted by Γ, that is, $\sim(\exists Z)\Gamma(X,Z).\equiv.(\exists S)\Gamma'(X,S)$. In the statements of these lemmas and the corresponding definitions we will omit references to $\Gamma=[E,H,\mathcal{K}]$; it remains the same throughout. In the following definition, $u\leq v<\omega$, c is a state of Γ, and C is a set of states of Γ (recall that the states of Γ are those of Z, that is, n-tuples of truth-values).

(2) $$S_u^{c,C}v \ : \ \{b;\exists Z[u,v].Zu=c \wedge (\forall t)_u^v H[Xt,Zt,Zt'] \wedge Zv=b \wedge \wedge \{Zu,\ldots,Zv-1\}=C\} .$$

The assertion about $Z[u,v]$ in this formula should be read: $Z[u,v]$ is C-<u>exact</u> X-<u>run</u> <u>of</u> H <u>from</u> c <u>to</u> b. Note that $S_u^{c,C}$ is an ω-sequence, whose states are subsets of the set of states of Γ, and which starts at u rather than 0. Thus, $S_u^{c,C}$ has 2^n states. We now pull these $m=n\cdot 2^n$ sequences together into one sequence S_u of m-tuples of sets of states of Γ.

(3) $$S_u v \ : \ < S_u^{c,C}v;\ c \text{ any state, } C \text{ a set of states of } \Gamma >$$

So the sequence S_u has still finitely many states, namely $g=2^{nm}$ of them. The reader will establish that each S_u can be defined by the recursion

(4) $$S_u u = e\,,\quad S_u t'=F[Xt,S_u t]\quad,\quad \text{for } u\leq t$$

whereby $e^{c,0} = \{c\}$, $e^{c,C} = 0$ if $C \neq 0$, and the $F^{c,C}$ are given by

$$b \in F^{c,C}[Xt,St] \;.\equiv.\; \bigvee_a [a \in C \wedge H[Xt,a,b] \wedge a \in S^{c,C}t \cup S^{c,C-a}t].$$

Such recursions are called finite-state recursions, or _finite automata_. This particular recursion $< e,F >$ is called _the_ (modified) _subset-construction_ on the transition part H of our system Γ, for finite ordinals. It is essential for our purpose, that one and the same recursion (4) works for all $u<\omega$ (uniformly). F will be used in making up the transition part of the system Γ' we are looking for. In fact, the conjunction over all $u<\omega$ of the recursion (4) would be a candidate for this transition part, were it not infinite state. We will now see that, in essence, there are but finitely many S_u. For this purpose define,

$u \approx v(t)$: $u,v \leq t \wedge S_u t = S_v t$ "u _and_ v _are merged at time_ t"

(5)

$u \approx v(-x)$: $(\exists t)^x u \approx v(t)$ "u _and_ v _merge before_ x"

Remark 3.2: For any X, the relation $\approx(-\omega)$ is an equivalence of finite index $\leq g$ on ω. Also $\approx(t)$ is an equivalence relation on t' of index $\leq g$, and $x \approx y(t)$, $t \leq z$ implies $x \approx y(z)$.

<u>Proof</u>: The second part is trivial, and so is the fact that $\approx(-\omega)$ is reflexive and symmetric. Suppose now $x \approx y(-\omega)$ and $y \approx z(-\omega)$. Then there are t_1 and t_2 such that $x \approx y(t_1)$ and $y \approx z(t_2)$. Hence, if $t = \max[t_1, t_2]$, $x \approx y(t)$ and $y \approx z(t)$. Therefore $x \approx z(t)$, and $x \approx z(-\omega)$. This shows that $\approx(-\omega)$ is transitive. Suppose now $x_0 < x_1 < \cdots < x_g$. Then two of these $(g+1)$ numbers must be in the relation $\approx(x_g)$, it being of index $\leq g$. Hence, these $x_i < x_j$ are also in the relation $\approx(-\omega)$. This argument shows that $\approx(-\omega)$ is of index $\leq g$. Q.E.D.

The merging relations thus show in which way just g of the infinitely many S_u are really different. In [2] the much more complicated merging relation $X[u,x)\sim X[v,y)$, defined by $S_u x = S_v y$ was used. Ramsey's theorem was used to show that this congruence on words is of finite index (strong version of remark 3.2). In [15] it is shown that Ramsey is actually not needed for showing that $\sim$ is of finite index. We will see that $\sim$ is not needed, except in its weak form (5). We will now make up formulas Ω which will play an essential role in the terminal part of the system Γ'. In the sequel s, s_0, s_1, s_2 will denote states of the sequences S_u. Note that the Ω, via (2) and (5), depend on the input X.

(6) $\qquad \Omega_{s_0,s} \;:\; (\exists w)(\exists^{\omega} y).S_0 y = s_0 \wedge S_w y = s \wedge y \approx w(-\omega)$

<u>Lemma</u> 3.3: For every X there are s_0, s such that $\Omega_{s_0,s}$.

<u>Proof</u>: Let X be arbitrary. By remark 3.2 the index of $\approx(-\omega)$ is finite. Hence, one of the equivalence classes must occur cofinal ω. That is, there is a w and an infinite (ω-cofinal) subset Q of ω, such that $Qy \supset y \approx w(-\omega)$. Now, $S_0 t$ and $S_w t$ take on but finitely many values. Hence, among the $y \in Q$ there are infinitely many at which S_0 and S_w take the same values, say s_0 and s. That is, there is an infinite $P \subseteq Q$ such that $Py \supset .S_0 y = s_0 \wedge S_w y = s$. As $P \subseteq Q$ we also have $Py \supset y \approx w(-\omega)$. Thus we have found s_0, s such that for some w and infinitely many y, $S_0 y = s_0 \wedge S_w y = s \wedge y \approx w(-\omega)$.

Q.E.D.

It is useful to have the following expressions for the Ω's.

(7) $$\Omega_{s_0,s} \;.\!\equiv\!.\; (\exists P).(\forall y)[Py \supset S_0y=s_0] \wedge$$
$$(\forall y)(\forall v)^y[Pv \wedge Py \supset S_vy=s] \wedge (\exists^\omega y)Py$$

<u>Proof</u>: Suppose first that $\Omega_{s_0,s}$. Then, by (6), there are a w and an infinite subset Q of ω such that,

$$Qv \supset S_0v=s_0 \wedge S_wv=s, \quad Qv \supset v\widetilde{\approx}w(-\omega).$$

We now define y_i, t_i by the following induction on $i<\omega$. ((μy) stands for "the first y")

$y_0 = (\mu y)Qy$ exists as Q infinite; Qy_0 and hence $y_0\widetilde{\approx}w(-\omega)$

$t_i = (\mu t)y_i\widetilde{\approx}w(t)$ exists as $y_i\widetilde{\approx}w(-\omega)$; $y_i\widetilde{\approx}w(t_i)$

$y_{i+1} = (\mu y).Qy \wedge t_i<y$ exists as Q infinite; Qy_{i+1} and hence $y_{i+1}\widetilde{\approx}w(-\omega)$, $t_i<y_{i+1}$

Note that $y_0\leqq t_0<y_1\leqq t_1<\dots$. As $y_i\widetilde{\approx}w(t_i)$ we also have $y_i\widetilde{\approx}w(y_j)$, for any $i<j$. As Qy_j we therefore have $S_{y_i}y_j = S_wy_j = s$, for any $i<j$. It now is clear that $P = \{y_i;\ i<\omega\}$ satisfies all the requirements on the right side of (7).

Suppose now that the right side of (7) holds. So we have an infinite P such that $Py \supset S_0 y = s_0$, $u<y \wedge Pu \wedge Py \supset S_u y = s$. Let w be the first member of P. So $y \neq w \wedge Py \supset S_w y = s$. Hence, $w \neq y < v \wedge Py \wedge Pv \supset y \approx w(v)$. As P is infinite this implies $w \neq y \wedge Py \supset y \approx w(-\omega)$. We now have $w \neq y \in P \supset S_0 y = s_0 \wedge$ $\wedge\ S_w y = s \wedge y \approx w(-\omega)$. Hence, by definition (6), $\Omega_{s_0,s}$.

Q.E.D.

We now are ready to prove the main lemma, which shows the relationship between the Ω's and the set of X accepted by the transition-system Γ. The following propositional expression B should be thought of as a set of pairs s_0, s of states of the S_u's.

$$(8) \qquad B[s_0,s] \quad : \quad \bigvee_{c,C,d,D} . E[c] \wedge \mathcal{K}[D] \wedge d \in s_0^{c,C} \cap s^{d,D}$$

Lemma 3.4: For any X, if $\Omega_{s_0,s}$ then $(\exists Z)\Gamma(X,Z) \equiv B[s_0,s]$. That is, if $\Omega_{s_0,s}$ then X is accepted by Γ just in case $< s_0, s >$ belongs to the set B.

Proof: Suppose $\Omega_{s_0,s}$ holds for an arbitrary input X. By (7) there is an infinite subset P of ω such that (a) $Py \supset S_0 y = s_0$ and (b) $v<y \wedge Pv \wedge Py \supset S_v y = s$.

Assume now $(\exists Z)\Gamma(X,Z)$. So by (1) there is a run Z such that $E[Z0]$, $(\forall t)H[Xt,Zt,Zt']$, and $\mathcal{K}[\sup Z]$. Let $c=Z0$, $D=\sup Z$. Then $E[c] \wedge \mathcal{K}[D]$. Furthermore, as P is infinite, we may assume (c) $v<y \wedge Pv \wedge Py \supset \{Zt; v \leq t < y\} = D$. (Else we could select an infinite subset Q of P which has this property; and being a subset of P, would still have the properties (a) and (b)). One of the values Zt, $t \in P$ must occur infinitely often; say this is d. Then clearly $d \in D$, as D=supZ. We may assume that (d) $Py \supset Zy=d$ (else we can select an infinite subset of P with this additional property).

Let y be the first member of P, let $C = \{Zt; 0 \leq t < y\}$. As $(\forall t)H(t)$, Z0=c, and Zy=d, it follows that Z[0,y] is a C-exact X-run of Γ. Hence, by the definition (2), $d \in S_0^{c,C}y$. By (a) we therefore have $d \in s_0^{c,C}$.

Let $v<y$ be any members of P. As $(\forall t)H(t)$ we have, by (d) and (c), that Z[v,y] is a D-exact X-run of Γ from d to d. Hence, by the definition (2), $d \in S_v^{d,D}y$. By (b) we therefore have $d \in s^{d,D}$. Going over these arguments we have found c,C,d,D such that $E[c] \wedge \mathcal{X}[D] \wedge d \in s_0^{c,C} \cap s^{d,D}$. Hence, by (8), $B[s_0,s]$. This argument shows $(\exists Z)\Gamma(X,Z) \supset B[s_0,s]$. It remains to prove the converse.

Assume then that $B[s_0,s]$. So we have c,C,d,D such that E[c], $\mathcal{X}[D]$, $d \in s_0^{c,C}$ and $d \in s^{d,D}$. Recall that we still have (a) and (b) from the first paragraph of this proof. These assert the existence of certain partial X-runs. We are to splice such to make up an X-run of Γ.

Let $y_0<y_1<y_2<\ldots$ be the enumeration of the infinite set P. By (a), $S_0y_0=s_0$, so by (3) we have $d \in S_0^{c,C}y_0$. Using the definition (2) this implies the existence of a C-exact X-run $Z[0,y_0]$ from c to d. Since E[c] we thus have

$$E[Z0] \wedge (\forall t)_0^{y_0} H[Xt,Zt,Zt'] \wedge Zy_0=d.$$

By (b) we have $S_{y_i} y_{i+1} = s$, so by (3) and $d \in s^{d,D}$, $d \in S^{d,D}_{y_i} y_{i+1}$. Using definition (2) this implies the existence of a D-exact X-run $Z[y_i, y_{i+1}]$ from d to d. So $Zy_i = d \wedge (\forall t)_{y_i}^{y_{i+1}} H[Xt, Zt, Zt'] \wedge$ $\wedge\ Zy_{i+1} = d$, and $\{Zt; y_i \leq t < y_{i+1}\} = D$.

As these partial runs $Z[0, y_0]$, $Z[y_0, y_1]$, $Z[y_1, y_2], \ldots$ have the same state d at the overlaps y_i, they may be spliced to form an ω-sequence $Z = Z[0, y_0) Z[y_0, y_1) Z[y_1, y_2) \ldots$. This Z clearly satisfies the initial condition E, and the transition-relation H. It is also clear that $\sup Z = D$, and as $\mathcal{X}[D]$, also the terminal condition $\mathcal{X}$ holds for Z. Hence by (1), $\Gamma(X, Z)$. We have thus shown $B[s_0, s] \supset (\exists Z) \Gamma(X, Z)$.

Q.E.D.

From Lemmas 3.4 and 3.5 we clearly obtain the formulas,

(9)

$$(\exists Z)\Gamma(X,Z) \quad .\equiv. \quad \bigvee_{\substack{s_0, s \\ B[s_0, s]}} \Omega_{s_0, s}$$

$$\sim(\exists Z)\Gamma(X,Z) \quad .\equiv. \quad \bigvee_{\substack{s_0, s \\ \sim B[s_0, s]}} \Omega_{s_0, s}$$

The second of these formulas is quite close to what we are looking for. What remains to be seen is that each $\Omega_{s_0, s}$ can be expressed as the behavior of a transition-system, and (7) almost does it.

Lemma 3.5: One can construct matrices $E'[Y0]$, $H'[Xt,Yt,Yt']$, and $L_{s_0,s}[Yt]$ such that for any X,

$$\Omega_{s_0,s} \equiv (\exists Y).E'(0) \wedge (\forall t)H'(t) \wedge (\exists^{\omega} t)L_{s_0,s}[Yt].$$

Proof: Recall the expression (7) for $\Omega_{s_0,s}$. Our Y will consist of the components P, Q, Y_0, Y_1, Y_2 whereby P, Q are set-variables and Y_0, Y_1, Y_2 range over sequences of states of the type s, that is sequences such as our S_u. The component P will correspond to the P in (7). Qx corresponds to $(\exists t)^x Pt$, and will be used as a device to capture the first member u_0 of P (namely u_0 is that t such that $Pt \wedge \widetilde{Qt}$). Y_0 corresponds to S_0, and therefore will be subjected to the recursion $< e,F >$ of (4), starting with $Y_0 0=e$. Due to (4) the condition $Py \supset S_0 y=s$ in (7) can now be replaced by $Pt \supset Y_0 t=s_0$. Y_1 and Y_2 will be used to simulate the condition $(\forall y)(\forall u)^y Pu \wedge Py \supset S_u y=s$, occurring in (7). The trick is to start Y_1, Y_2 off at the first $u \in P$, with the value e, and from there on use the recursion F to find $Y_1 t'$ and $Y_2 t'$, with the exception that at each $y \in P$, $Y_2 y$ is reset to e. Because of (4) it then is clear that the condition $(\forall y)(\forall u)^y Pu \wedge Py \supset S_u y=s$ can be replaced by $Pt' \wedge t' \neq u_0 \supset Y_1 t'=s \wedge F[Xt,Y_2 t]=s$. Formally stated all this comes to,

$$\Omega_{s_0,s} \equiv (\exists PQY_0Y_1Y_2).E'(0) \wedge (\forall t)H'(t) \wedge \Phi s_0,s$$

whereby E', H', $\Phi_{s_0,s}$ are the conjunctions of the following items.

$Y_0 0 = e$		$\tilde{Q}0$	initial condition E'(0)
	$Y_0 t' = F[Xt, Y_0 t]$	$Qt' \equiv [Qt \vee Pt]$	transition condition H'(t)
$\tilde{Q}t \wedge Pt \supset Y_1 t = e$	$Qt' \supset Y_1 t' = F[Xt, Y_1 t]$		
$Pt \supset Y_2 t = e$	$\tilde{P}t' \wedge Qt' \supset Y_2 t' = F[Xt, Y_2 t]$	$Pt' \wedge Qt' \supset F[Xt, Y_2 t] = Y_1 t'$	
	$(\exists^{\omega} t).Pt \wedge Y_0 t = s_0 \wedge Y_1 t = s$		terminal condition $\Phi_{s_0,s}$

Actually, starting from $\Omega_{s_0,s}$ given by (7), the P will uniquely determine Q, Y_0, Y_1, Y_2 satisfying E'(0), $(\forall t)H'(t)$ (except for an initial segment of Y_1 and Y_2). These P, Y_0, Y_1, Y_2 then will satisfy the output conditions $(\exists^{\omega} t)Pt$, $(\forall t)Pt \supset Y_0 t = s_0$, and $(\forall t)Pt \supset Y_1 t = s$. From these $\Phi_{s_0,s}$ follows. The other way around, assume P, Q, Y_0, Y_1, Y_2 are such that E'(0), $(\forall t)H'(t)$, $\Phi_{s_0,s}$. Then we have an infinite subset $P_1 \subseteq P$ such that all elements $t \in P_1$ satisfy $Y_0 t = s_0$, $Y_1 t = s$. Using E'(0), $(\forall t)H'(t)$ one shows, by (4), that P_1 satisfies the right side of (7), and hence $\Omega_{s_0,s}$.

Q.E.D.

<u>Complementation lemma</u> 3.6: To every transition-system $\Gamma(X,Z)$, of form (1), one can construct a transition-system $\Gamma'(X,Y)$ of the same form, such that $\sim(\exists Z)\Gamma(X,Z) \equiv (\exists Y)\Gamma'(X,Y)$. That is, the set of ω-sequences X accepted by the new system Γ' is the complement of that accepted by the original system Γ. In fact, the terminal condition $\mathcal{X}'[\sup Y]$ of Γ' is of the simple

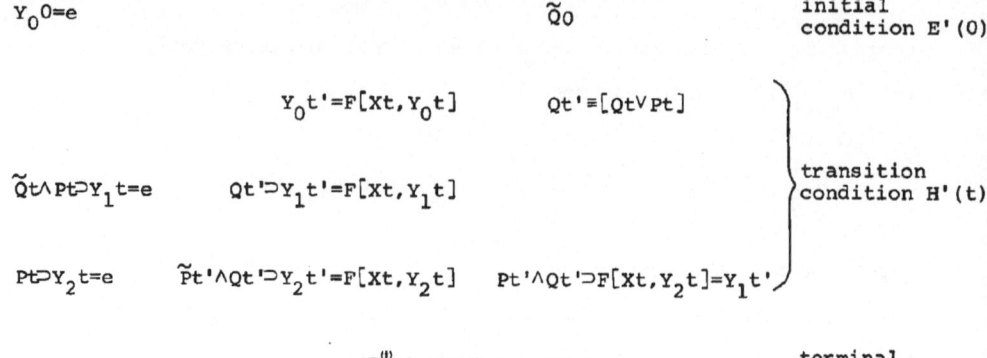

form $(\exists^{\omega} t)L'[Yt]$.

Proof: Starting from Γ we use Lemmas 3.3 and 3.4 to obtain the equivalence (9) of $\sim(\exists Z)\Gamma(X,Z)$ to $\bigvee_{s_0,s} \Omega_{s_0,s}$. Here and in the sequel $< s_0, s >$ ranges over the finite set $\sim B$. By Lemma 3.5 each $\Omega_{s_0,s}$ is of form $(\exists Y).E'(0) \wedge (\forall t)H'(t) \wedge (\exists^{\omega} t)L_{s_0,s}[Yt]$. As s_0, s do not occur in E' and H', the disjunction $\bigvee_{s_0,s}$ can be moved through $(\exists Y)$ and in front of $(\exists^{\omega} t)L_{s_0,s}(t)$. As a finite union of sets is infinite just in case one of the terms is infinite, the disjunction can also be moved through $(\exists^{\omega} t)$. Hence we have

$$\sim(\exists Z)\Gamma(X,Z) \ .\!\equiv.\ (\exists Y).E'(0) \wedge (\forall t)H'(t) \wedge (\exists^{\omega} t)L'[Yt] ,$$

whereby $L' = \bigvee_{s_0,s} L_{s_0,s}$. Note that $(\exists^{\omega} t)L'[Yt]$ states $L' \cap \sup Y \neq 0$. So $\Gamma' = [E',H',\mathcal{X}']$, where $\mathcal{X}' = \{D; L' \cap D \neq 0\}$, is the required system.

Q.E.D.

The new system Γ' which we constructed has $2^2 \cdot g^3 = 2^{2+3nm}$ states. These states may be coded as $(2 + 3nm)$-tuples of truth-values, so the runs Y of Γ' become $(2 + 3nm)$-tuples of monadic predicates. As we have pointed out in Section 2, the complementation Lemma 3.6 at once yields the following information about definability.

<u>Definability Theorem</u> 3.7: Every MT-formula $\Sigma(X)$ is equivalent, in $[\omega,<]$, to a formula of form

$$(\exists Z).A[Z0] \wedge (\forall t)B[Xt,Zt,Zt'] \wedge (\exists^{\omega}t)C[Zt].$$

I.e., a n-ary relation R on subsets of ω is MT-definable in $[\omega,<]$ just in case it is the set of accepted input-sequences $X = X_1,\ldots,X_n$ of a finite transition system $[A,B,\mathcal{C}]$.

As there is a very nice intuitive background to the working of a transition-system, this theorem yields rather concrete information on definability in $MT[\omega,<]$. Some consequences on definability are discussed in [15]. A stronger form of our definability theorem 3.7 will be obtained by using the subset-construction rather than the complementation lemma. We now state the following special case of 3.7.

<u>Theorem</u> 3.8 (<u>normal form of sentences</u>): To every MT-sentence Σ one can construct an input-free transition-system $\Gamma = E[Z0]$, $H[Zt,Zt']$, $C[Zt]$, such that Σ holds in $[\omega,<]$ just in case Γ admits a run, i.e., just in case

$$(\exists Z).E[Z0] \wedge (\forall t)H[Zt,Zt'] \wedge (\exists^{\omega}t)C[Zt] .$$

<u>Item</u> 3: <u>The Decision-method</u>. We will now show that, given the input-free system $\Gamma = [E,H,\mathcal{K}]$, one can decide whether or not Γ admits a run. Because of Theorem 3.8 this will at once yield the decidability of $MT[\omega,<]$. The input-free system Γ looks as

follows.

$$(1') \qquad \Gamma(Z) \quad : \quad E[Z0] \wedge (\forall t)H[Zt,Zt'] \wedge \mathcal{K}[\sup Z] \quad .$$

It is easy to see that such a system has a run, only if it has an ultimately periodic run. Furthermore one can put a bound on phase and period. It now is clear that one can decide whether or not $(\exists Z)\Gamma(Z)$. It is interesting (and useful at ω_1) to note that all this is actually contained in our basic Lemma 3.4. This is seen as follows.

Remark 3.9: Given the input-free system one can construct s such that $\Omega_{s,s}$.

Proof: From the definition (2), in the case where X is not present (i.e., has just one state), it is obvious that we have (a) $S_0t = S_u(u+t)$, and (b) $S_0x = S_0y.\supset.S_0(x+t) = S_0(y+t)$. As S_0 has finitely many states, we can use the recursion (4), $S_00=e$, $S_0t'=F[S_0t]$, to find $u<u+v$ such that $S_0u = S_0(u+v)$. Using (a), this yields $S_v(u+v) = S_0(u+v)$, and therefore $v \approx 0(-\omega)$. Iterating this procedure yields (c) $vt \approx 0(-\omega)$. Using the recursion (4), we can next find $i<i+j\leq g$, such that $S_0vi = S_0v(i+j)$. Call this state s. By (b) we have $S_0v(i+jt) = s$. Together with (c) this implies $(\exists^\omega y).y \approx 0(-\omega) \wedge S_0y=s$. Therefore, by definition (6), $\Omega_{s,s}$.

Q.E.D.

Theorem 3.10: The monadic second order theory of $[\omega,<]$ is decidable.

Proof: Given the sentence Σ, construct the input-free system Γ of Theorem 3.8. Now find the s of Remark 3.9. As $\Omega_{s,s}$, we know by Lemma 3.4, that Σ, i.e., $(\exists Z)\Gamma(Z)$ holds just in case $B[s,s]$. From its definition (8), one can clearly evaluate this. Q.E.D.

Item 2': The subset-construction at ω. Our proof of the complementation lemma depends heavily on the availability of a uniform subset-construction (4) which works for all $\alpha<\omega$. To obtain a complementation lemma at ω_1 we will therefore need to extend this subset-construction uniformly to all $\alpha<\omega_1$. We will show here, using ideas of McNaughton [13], how this works for $\alpha=\omega$. Once this first limit is carefully handled, the extension to all ω-accessible limits will be easy.

Note that from Lemmas 3.3 and 3.4 we obtained the important rough form (9) of the complementation lemma. Only in Lemma 3.5 we used the subset-construction (4), to put the disjunction $\bigvee \Phi_{s_0,s}$ into the required form. The result was a non-deterministic transition system; non-deterministic because the reset-points $y\in P$ for the recursion on the component Y_2 were not uniquely determined. What will be done now may be viewed as a remedy for this matter; we are looking for a deterministic version of Lemma 3.5.

So, the system Γ will still be the one given by (1), and all the ideas and results (except 3.5), used in the proof of the complementation lemma, will be used also in the deterministic version. We begin by restating these facts in a more concise form. The formula Ω is defined by,

$$(10) \qquad \Omega \; : \; (\exists w)(\exists^{\omega} y).y \approx w(-\omega) \wedge B[S_0 y, S_w y] \; .$$

<u>Lemma</u> 3.11: For any input X, $(\exists Z)\Gamma(X,Z)$ holds just in case the sequences S_u , defined by the recursion (4) from X, satisfies the terminal condition Ω.

<u>Proof</u>: As the finite subsets of ω form an ideal, a disjunction can be moved through the quantifier $(\exists^{\omega} y)$. Also note that $\bigvee_{s_0,s} . B[s_0,s] \wedge S_0 y = s_0 \wedge S_w y = s$ just means $B[S_0 y, S_w y]$. Hence, by (6) and (10), Ω is equivalent to $\bigvee_{s_0,s} \Omega_{s_0,s}$, whereby the index s_0,s ranges over B. Now our lemma becomes just a restatement of formula (9).

Q.E.D.

Thus, we already possess a deterministic system for $(\exists Z)\Gamma(X,Z)$. The tasks are,

A: to simulate the recursion (4) by one which is finite-state.

B: to put the terminal condition Ω into sup-form.

As for A we have the hint in Remark 3.2: at any time t at most g of the S_u's are different. The idea is to shut off S_u

as soon as u merges with an earlier v, and from there on use v as a representative for u. This focuses our attention on the following notion, which will be important also for task B.

(11) $u \sim v(t) : u \approx v(t) \wedge u \not\approx v(-t)$ "u and v merge at time t"

We will postpone the details on A, as the real problem and work is in task B. Note that in the definition (10) of our terminal condition Ω a quantifier $(\exists t)$ is hidden in the part $y \approx w(-\omega)$, as this means $(\exists t) y \sim w(t)$. Furthermore, because of the way we want to handle A, the quantifier $(\exists^{\omega} y)$ in Ω is bad. The following new form of Ω is essential.

(12) $$\Omega .\equiv. (\exists w)(\exists^{\omega} t)(\exists y).y \sim w(t) \wedge B[S_0 y, S_w y] .$$

Proof: Suppose first Ω. So by (10) we have a w and an infinite set P such that (a) $Py \supset (\exists t) y \approx w(t)$, and (b) $Py \supset B(y)$. As in the proof of (7) we now define y_i, t_i for all $i<\omega$, by the following induction.

$y_0 = (\mu y) Py$ exists as P is infinite; Py_0

$t_i = (\mu t) y_i \approx w(t)$ exists as Py_i and (a); $y_i \leqq t_i$ and $y_i \sim w(t_i)$

$y_{i+1} = (\mu y).Py \wedge t_i < y$ exists as P is infinite; Py_i and $t_i < y_{i+1}$

As $y_i \leqq t_i < y_{i+1}$ the set $\{t_i ; i<\omega\}$ is infinite. As Py_i we have by (b), $B(y_i)$. Hence $(\exists^{\omega} t)(\exists y).y \sim w(t) \wedge B(y)$, which proves the implication (12) from left to right.

Suppose now that the right side holds. So we have w, an infinite set T, and a function y_t such that (c) $Tt \supset y_t \sim w(t)$, and (d) $Tt \supset B(y_t)$. Now let z be arbitrary and define $A = \{t; Tt \wedge y_t \leq z \leq t\}$. Suppose $t_1 \leq t_2$, are two members of A, and $y_{t_1} \approx y_{t_2}(z)$. As $z \leq t_1$ we have $y_{t_1} \approx y_{t_2}(t_1)$, and by (c), $y_{t_1} \approx w(t_1)$. Therefore $y_{t_2} \approx w(t_1)$. But by (c), $y_{t_2} \sim w(t_2)$. Hence $t_2 \leq t_1$, and as $t_1 \leq t_2$, $t_1 = t_2$. This argument shows that $y_{t_1} \approx y_{t_2}(z)$ is the equality relation on A. But this relation is also an equivalence of finite index (since $\approx(z)$ is). Therefore A is finite. As T is infinite we thus have shown, $(\forall z)(\exists t)[Tt \wedge z < y_t]$. Hence by (c, d) $(\forall z)(\exists y)_z[y \approx w(-\omega) \wedge B(y)]$. By (10) this implies Ω. Q.E.D.

This new form of Ω is still not what we need, as there is reference not just to t, but also to a y which may be in the remote past of t. To remedy this, McNaughton invented the following piggyback device, which carries the needed information at y along to t.

(13)

$r_y t$:	$(\mu z) z \approx y(-t)$	"the representative of y at t"
Ut :	the sequence in natural order, of all $r_y t$, $y<t$	"the vector of indices active at t"
$u \leftarrow v(t)$:	v in Ut, and u the first in Ut such that $u \approx v(t)$	"in Ut', u replaces v"
$Q_w t$:	$\{r_y t; w \leq y < t \wedge B[S_0 y, S_w y]\}$	"representatives at t of past hits"

The task is first to show the new form of Ω (see 14), and second to find recursions for the newly added components U and Q_w (see Lemma 3.12). While the proofs of these items are really quite obvious, due to the nice intuition going with (13), the details are tricky. We begin with a list of obvious consequences of (13).

(i) $y<t \wedge v=r_y t .\supset. v\leqq y<t \wedge v \approx y(-t)$

(ii) $w\leftarrow w(t) \wedge r_y t \approx w(t) .\supset. w\leftarrow r_y t(t)$

(iii) $w=r_y x \wedge w<t<x .\supset. w\leftarrow w(t)$

(iv) $y \approx w(-t) .\supset. r_y t=r_w t$

(j) $w\leftarrow w(t) .\supset. w<t \wedge w=r_w t$

(jj) $t\neq v\in Ut' .\supset. v\leftarrow v(t)$

(jjj) $v\leftarrow v(t) .\supset. v\in Ut \wedge v\in Ut'$

(jw) $r_t t'=t,\ t\in Ut',\ w<t \supset w \not\approx t(t)$

(v) $y<t .\supset. r_y t' \leftarrow r_y t(t)$

(vi) $v=r_y t \wedge t<x \wedge v\in Ux .\supset. v=r_y x$

(w) $u\leftarrow v(t) \wedge w\leftarrow v(t) .\supset. u=w$

In the case of (jw) note that the first two items are a consequence of the third. To see the third item note that $w \approx t(t)$ implies $S_w^{c,0}\ t = S_t^{c,0}\ t = \{c\}$, and $S_w^{c,0}\ t = \{c\}$ is impossible for $w<t$.

We are now ready to show the new form of our terminal condition.

$$(14)\quad \Omega .\equiv. (\exists w).(\forall^{\omega} t) w\leftarrow w(t) \wedge (\exists^{\omega} t)(\exists v)[v\neq w\leftarrow v(t) \wedge v\in Q_w t]$$

<u>Proof</u>: Suppose first that Ω. By (12) there are w and infinitely many t_i, y_i such that (a) $y_i \sim w(t_i)$, and $B(y_i)$. We may assume that $w<t_i$, and that $w=r_w\omega$ is the first in its class modulo $\approx(-\omega)$. Then by (iii) we have (b), $(\forall t)_w, w\nvdash w(t)$, and in particular $w\nvdash w(t_i)$. As $w<t_i$ we have by (j) $t_i \not\approx w(t_i)$, and by (a), $y_i<t_i$. Therefore $v_i=r_{y_i}t_i$ exists, and by (i), $v_i \leq y_i<t_i$ and $y_i \approx v_i(-t_i)$. By (a) this yields, $v_i \neq w$ and $v_i \approx w(t_i)$. Hence by (ii) and $w\nvdash w(t_i)$, we have $w\nvdash v_i(t_i)$. So by (13), $w \leq v_i$, and as $v_i \leq y_i<t_i$, we have $w \leq y_i<t_i$. As $B(y_i)$ we use (13) to obtain $v_i \in Q_w t_i$. Because of (b), and as there are infinitely many t_i, $v_i \neq w \nvdash v_i(t_i) \wedge v_i \in Q_w t_i$, we get the right side of (14).

Assume now the right side of (14). So we have w and infinitely many t_i, v_i such that $w\nvdash w(t_i)$, $v_i \neq w$, $w\nvdash v_i(t_i)$, and $v_i \in Q_w t_i$. By (13) we conclude the existence of y_i such that $v_i=r_{y_i}t_i$, $w \leq y_i<t_i$, and $B(y_i)$. Hence $v_i \approx y_i(t_i)$ and $w \approx v_i(t_i)$, so $y_i \approx w(t_i)$. As $w\nvdash w(t_i)$ and $w<t_i$ we use (j) to get $r_w t_i=w\neq v_i=r_{y_i}t_i$. Hence by (iv), $y_i \not\approx w(-t_i)$. So $y_i \sim w(t_i)$. As $B(y_i)$, and there are infinitely many t_i, we use (12) to obtain Ω.

Q.E.D.

It remains to replace the definition (13) of U and Q by recursions, so they can be used to carry on to t, the information about earlier hits, by reference to the immediate past, only.

<u>Lemma</u> 3.12: The sequences U, Q_w defined by (13) satisfy the following recursion:

$$U0 = \text{empty} \qquad\qquad Q_w w = 0$$

$$Ut' = \text{remove from } Ut \text{ all } v \not\leftarrow v(t), \text{ and add } t \text{ at end}$$

$$Q_w t' = \{v; (\exists p)[v \leftarrow p(t) \wedge p \in Q_w t] \vee [v = t \wedge B[S_0 t, S_w t]]\}$$

<u>Proof</u>: The initial equations are trivial, by (13). The equation for Ut' follows from (jj, jjj, jw). We are left to show the equation for $Q_w t'$, which comes to showing:

(a) $v \in Q_w t' . \supset . (\exists p)[v \leftarrow p(t) \wedge p \in Q_w t] \vee [v = t \wedge B(t)]$

(b) $v \leftarrow p(t) \wedge p \in Q_w t . \supset . v \in Q_w t'$

(c) $B(t) . \supset . t \in Q_w t'$

Suppose $v \in Q_w t'$. Then by (13) there is a y such that $v = r_y t'$ and $w \leq y \leq t$ and $B(y)$. In case $y = t$, the second disjunct in (a) holds. We may therefore assume $w \leq y < t$. Let $p = r_y t$. By (v) it follows, $v \leftarrow p(t)$, and by (13) and $B(y)$, $p \in Q_w t$. Hence the second disjunct in (a) holds. This proves (a).

Suppose now $v \leftarrow p(t)$, $p \in Q_w t$. By (13) there is a y such that $w \leq y < t$, $r_y t = p$, and $B(y)$. By (v) we have $r_y t' \leftarrow p(t)$, and therefore by (w), $v = r_y t'$. As $B(y)$ and $w \leq y < t$ this implies, by (13), $v \in Q_w t'$. This proves (b). Suppose now $B(t)$. By (jw) we have $r_t t' = t$. Hence by (13), $t \in Q_w t'$. This proves (c).

Q.E.D.

Adding the recursions for U, Q_w (of Lemma 3.12) to those in Lemma 3.11, and interpreting the terminal condition Ω in the new form (14), answers our problem of putting $(\exists Z)\Gamma(X,Z)$ into deterministic form with sup-condition. At any time t, at most $(g+1)$ of the indices u are active, i.e., occur in Ut. Furthermore, we may shut off or reset S_u and Q_u as soon as u becomes inactive because this will not affect later active parts of the recursion, and because only active u's enter into the terminal condition Ω. It is also clear intuitively that the following trick will not essentially change the situation: At any time t, let $\zeta(t)$ be the first v which is not in Ut. The trick consists in letting $\zeta(t)$ take over the function of t, from t' on. In particular, at time t' we add $\zeta(t)$, rather than t, as last entry in the sequence Ut'. More precisely this leads to the following recursion.

$V0$ = empty $\qquad\qquad$ Vt' = remove from Vt all $i \not\gamma i(t)$ and add $\zeta(t)$ at end

$$(15) \quad S_i t = \begin{cases} e & \text{if } i \text{ not in } Vt \\ \mathscr{I}_i(t) & \text{if } i \text{ in } Vt \end{cases} \qquad Q_k t = \begin{cases} 0 & \text{if } k \text{ not in } Vt \\ \mathscr{Q}_k(t) & \text{if } k \text{ in } Vt \end{cases}$$

$$\bigvee_k .(\forall^\omega t) k \leftarrow k(t) \wedge (\exists^\omega t) \bigvee_j [j \neq k \wedge k \leftarrow j(t) \wedge j \in Q_k t] \quad \text{the terminal condition}$$

Here the indices i, j, h, k range over $\{0,\ldots,g\}$, and the following abbreviations are used: $e^{c,0} = \{c\}$, $e^{c,C} = 0$ if $C \neq 0$, and

$S_i t$: $< S_i^{c,C} t; c$ a state of Γ, C a set of states of $\Gamma >$

$i \leftarrow j(t)$: j in Vt, i first in Vt such that $S_i t = S_j t$
"i takes over j at time t"

$\zeta(t)$: the first $i \leq g$, i not in Vt "the index activated at t'"

$\mathcal{S}_i^{c,C}(t')$: $\{b; \bigvee_{a \in C} .H[Xt,a,b] \wedge a \in (S_i^{c,C} t \cup S_i^{c,C-a} t)\}$
"the tentative value of $S_i^{c,C} t'$ "

$\mathcal{Q}_k(t')$: $\{j; \bigvee_h [j \leftarrow h(t) \wedge h \in Q_k t] \vee [j = \zeta(t) \wedge B[S_0 t, S_k t]]\}$
"the tentative value of $Q_k t'$ "

With the hints given about the relation between t and $\zeta(t)$, the reader will verify just in which sense the recursion (15) for S_i, Q_k, V simulate those for S_u, Q_w, U (in Lemmas 3.11 and 3.12). In this sense the terminal condition in (15) corresponds to the form (14) of the conditions Ω. Hence, by Lemma 3.11, $(\exists Z)\Gamma(X,Z)$ holds just in case the recursion (15) applied to X determines such S_i, Q_k, V which satisfy the terminal condition in (15). We will state and formally prove this in Lemma 3.13. Note that the recursion (15) is on finitely many components S_i, Q_k, V, each taking finitely many states. Namely, the states of S_i are the familiar objects s (g of them), the states of Q_k are subsets of $\{0,\ldots,g\}$, and the states of V are vectors with entries $i \leq g$, and no repetitions. By the Remark 1.2 in Section 1, the output-condition in (15) can be stated as a sup-condition on the total sequence Y, consisting of the components

S_i, Q_k, V. Thus, (15) is a transition-system of the following form,

(16) $\Delta(X,Y)$: $Y0=a \wedge (\forall t)Yt'=F[Xt,Yt] \wedge \mathcal{J}[\sup Y]$

whereby Y takes on finitely many states, F maps pairs of input states and states to states, and $\mathcal{J}$ is a set of sets of states. We call such a Δ a <u>deterministic transition-system</u>, or <u>finite automaton</u>. It should be noted that the states of the automaton Δ may be coded as m-vectors of truth-values. Y thus becomes a vector of m set-variables, a becomes an m-vector of truth-values, F and $\mathcal{J}$ can be expressed as matrices (i.e., Boolean formulas) in the indicated atomic parts.

<u>Lemma</u> 3.13 (<u>the subset-construction</u>): To every transition-system $\Gamma(X,Z)$ of form (1), one can make up a deterministic system $\Delta(X,Y)$ of form (16), such that for all X, $(\exists Z)\Gamma(X,Z)\equiv(\exists Y)\Delta(X,Y)$. In fact, the subset-construction (15) on Γ will yield such a Δ. That is, for any input X, Γ accepts X just in case the unique S, Q, V determined by the recursions (15) satisfies the terminal condition.

<u>Proof</u>: We ask the reader to write down the formulas $(\overline{15})$, obtained from (15) by making the following replacements: i,j,k,h by u,v,w,p; $S,\mathcal{J},Q,\mathcal{Q}$ by $\overline{S}$, $\overline{\mathcal{J}}$, $\overline{Q}$, $\overline{\mathcal{Q}}$; V by U; $\leftarrow$ by $<=$; $\zeta(t)$ by t . Now let X be arbitrary, let $\overline{S}_u$, U, $\overline{Q}_w$ be the

sequences defined by the recursion ($\overline{15}$), and let S_i, V, Q_k be the sequences defined by the recursion (15). Note that by (14) the terminal condition in ($\overline{15}$) is just Ω. Hence by Lemmas 3.11 and 3.12 we have:

(a) $(\exists Z)\Gamma(X,Z)$ if and only if $\overline{S}_u$, U, $\overline{Q}_w$ satisfy the terminal condition in ($\overline{15}$). Suppose now we have a function $u_i t$ which, at any time t, establishes a one-one relation between the vectors Vt and Ut of active indices, such that for active i, $S_i t=\overline{S}_{u_i t} t$ and $Q_i t=\overline{Q}_{u_i t} t$. I.e.,

(b_1) $i \text{ in } Vt.\supset.u_i t \text{ defined, in } Ut$

(b_2) $u \text{ in } Ut.\supset.(\exists i)[i \text{ in } Vt \wedge u_i t=u]$

(b_3) $i \text{ before } j \text{ in } Vt.\supset.u_i t<u_j t$

(b_4) $i \text{ in } Vt \wedge u=u_i t.\supset.S_i t=\overline{S}_u t$

(b_5) $j \text{ in } Vt.\supset.i\leftarrow j(t)\equiv u_i t\Leftarrow u_j t(t)$

(b_6) $i \text{ in } Vt \wedge u=u_i t.\supset.j\in Q_i t\equiv u_j t\in\overline{Q}_u t$

As only active indices enter into the terminal conditions, it is easy to see that, using this $u_i t$, one has: The sequences $\overline{S}_u$, U, $\overline{Q}_w$ satisfy the terminal condition in ($\overline{15}$) just in case the sequences S_i, V, Q_k satisfy the terminal condition in (15). Together with (a) this will prove our Lemma. Thus, it remains only to set up the function $u_i t$, and to prove that it has the properties (b). We define $u_i t$ inductively

(c) $u_i t'=u_i t, \text{ if } i\leftarrow i(t) \quad u_i t'=t, \text{ if } i=\zeta(t)$.

That (b) hold at $t=0$ is trivial, because $V0$ and $U0$ are empty. We now carry out the inductive step from t to t'.

b_1: Suppose i is in Vt'. Then by (15) we have the following two cases.

<u>Case</u> $i=\zeta(t)$: By (c) we have $u_i t' = t$. Hence by $(\overline{15})$, $u_i t'$ is in Ut'.

<u>Case</u> $i \neq i(t)$: Then i is in Vt. By inductive assumption $u_i t$ is defined in Ut, and $u_i t \lessdot u_i t(t)$. Hence by $(\overline{15})$, $u_i t$ is in Ut'. By (c), $u_i t' = u_i t$. Hence also in this case, $u_i t'$ is defined and in Ut'.

b_2: Suppose u is in Ut'. By $(\overline{15})$ we have the following two cases.

<u>Case</u> $u=t$: Let $i=\zeta(t)$. By (15) and (c), i in Vt' and $u_i t' = u$.

<u>Case</u> $u \lessdot u(t)$: Then u is in Ut. By inductive assumption there is an i in Vt, $u=u_i t$. Hence $u_i t \lessdot u_i t(t)$, and by inductive assumption $i \neq i(t)$. Therefore, by (15) and (c), i is in Vt' and $u_i t' = u_i t = u$.

b_3: Suppose i is before j in Vt. By (15) we have the following two cases.

<u>Case</u> $i \neq i(t)$, $j=\zeta(t)$: By (c), $u_i t' = u_i t$ in Ut and $u_j t' = t$. By $(\overline{15})$ it follows that $u_i t' < u_j t'$.

<u>Case</u> $i \neq i(t)$, $j \neq j(t)$, i before j in Vt: By inductive assumption, $u_i t < u_j t$, and by (c), $u_i t' = u_i t$ and $u_j t' = u_j t$. Hence also in this case $u_i t' < u_j t'$.

b_4: Suppose i is in Vt' and $u=u_i t'$. By (15) we have two cases.

Case $i=\zeta(t)$: Then i is not in Vt, and by (15) $S_i t=e$. Now by (c) $u=t$, and therefore by $(\overline{15})$, $\overline{S}_u t=e$. Hence by (15) and $(\overline{15})$ we have $S_i t'=\mathcal{S}_i(t')=\overline{\mathcal{S}}_u(t')=\overline{S}_u\ t'$.

<u>Case</u> $i\leftarrow i(t)$, i in Vt: By (c) we have $u=u_i t$. By inductive assumption $S_i t=\overline{S}_u t$, and $u<=u(t)$. By (15) and $(\overline{15})$ it follows, $S_i t'=\mathcal{S}_i(t')$ and $\overline{S}_u t'=\overline{\mathcal{S}}_u(t')$, and because $S_i t=\overline{S}_u t$, $\mathcal{S}_i(t')=\overline{\mathcal{S}}_u(t')$. Therefore also in the second case, $S_i t'=\overline{S}_u t'$.

b_5: We show that b_5 is a consequence of b_1, b_2, b_3, b_4. Suppose j is in Vt. $i\leftarrow j(t)$ is equivalent by (15) to, i is first in Vt such that $S_i t=S_j t$, is equivalent by b_1, b_2, b_3, b_4 to, $u_i t$ is first in Ut such that $\overline{S}_{u_i} t=\overline{S}_{u_j} t'$ is equivalent by $(\overline{15})$ to, $u_i t<=u_j t(t)$.

b_6: Suppose k is in Vt' and $w=u_k t'$. In case $k=\zeta(t)$ we have k is not in Vt and $w=t$. Therefore, by (15) $Q_k t=0$, and by $(\overline{15})$ $Q_w t=0$. Hence, by (15) and $(\overline{15})$, $Q_k t'=\mathcal{Q}_k(t')=\overline{\mathcal{Q}}_w(t')=\overline{Q}_w t'$, so b_6 holds. We therefore may assume now that $k\leftarrow k(t)$ and k is in Vt. By using (c) and the inductive assumption we get $w<=w(t)$, and w is in Ut. By (15) and $(\overline{15})$ it follows that $Q_k t'=\mathcal{Q}_k(t')$ and $\overline{Q}_w t'=\overline{\mathcal{Q}}_w(t')$. It therefore remains to show, $j\in\mathcal{Q}_k(t')\equiv u_j t'\in\overline{\mathcal{Q}}_w(t')$. This proof goes as follows:

$$j\in\mathcal{Q}_k(t') \overset{(15)}{\longleftrightarrow} [j=\zeta(t)\wedge B[S_0 t,S_k t]\vee\bigvee_h[j\leftarrow h(t)\wedge h\in Q_k t]$$

$$\updownarrow(c)\qquad\updownarrow\text{in. ass.}\qquad\qquad\updownarrow\text{ind. assp.}$$

$$[u_j t'=t\wedge B[\overline{S}_0 t,\overline{S}_w t]\vee\bigvee_h[u_j t<=u_h t(t)\wedge u_h t\in\overline{Q}_w t]\overset{(\overline{15})}{\underset{*}{\longleftrightarrow}} u_j t'\in\overline{\mathcal{Q}}_w(t')$$

The * means that $u_j t' = u_j t$ is used, and also the remark $v \in \overline{Q}_w t \supset v$ in Ut. This is easily shown by induction on t. It is needed because $u_j t' \in \overline{\mathfrak{Z}}_w(t')$ does not directly yield the h in the second disjunct. We only get a v such that $u_j t <= v(t) \wedge v \in \overline{Q}_w t$. * is now used to infer $v \in Ut$, and hence by inductive assumption, $v = u_h t$ for some h.

Q.E.D.

As we have remarked in Section 2, the subset-construction 3.13 implies the complementation lemma 3.6, and hence Theorems 3.7 and 3.10. Now, it is clear that one does not want to use the subset-construction in a decision-method for $MT[\omega,<]$. The non-deterministic construction, in the proof of Lemma 3.5, is much more economical. However, the subset-construction yields much better information on definability, and the form of sentences. We clearly have the following strong version of Theorems 3.7 and 3.8.

Definability Theorem 3.7': To every MT-formula $\Sigma(X)$ in the primitive $<$ one can construct a finite automaton $A = [a,F,\mathfrak{F}]$ such that, for any X, $\Sigma(X)$ holds in $[\omega,<]$ if and only if A accepts X.

Theorem 3.8' (normal form for sentences): To every MT-sentence Σ in $<$ one can construct an input-free finite automaton A, such that Σ holds in $[\omega,<]$ if and only if the (unique) run of A satisfies the terminal condition.

The nicest way to realize the strength of 3.7' is to note that a finite state recursion defines a continuous function $Y=fX$ from ω-sequences to ω-sequences, and that $\mathcal{S}[\sup Y]$ defines a set G of sequences which belongs to the Boolean algebra $B(F_\sigma)$ over the F_σ's. Hence, the set $f^{-1}(G)$ consisting of all X accepted by a finite automaton $[f,G]$ still is in $B(F_\sigma)$. From 3.7' we therefore see that only $B(F_\sigma)$'s can be defined in $MT[\omega,<]$. (For more details, see [6].) From 3.7 one only sees that every definable set is (Souslin and co-Souslin and therefore) Borel; but 3.7 does not yield any bound, in the Borel-hierarchy, for definable sets.

Another important application, where 3,7' is essential, occurs in [5], where it is used to extend a result on finite-state games to a solvability algorithm for $MT[\omega,<]$. In the next section we will see that the deterministic form of the complementation lemma extends to arbitrary countable ordinals. In contrast, it is not clear how one extends the non-deterministic construction in Lemma 3.5 to ω^ω.

4. The subset-construction extended up to ω_1

We will here collect the bonus for having carefully analyzed, and put into rigorous form, McNaughton's ideas. Namely, the proof of the complementation lemma in Section 3, up to Lemma 3.5, extends verbatim to any ω_0-accessible limit α. As we do not, at this place, have an extension of the subset-construction (formulas (4) in Section 3), we do not see how to extend the non-deterministic construction in the proof of Lemma 3.5. We therefore have to substitute for it the much more intricate deterministic construction, Lemma 3.11 through 3.13. But the extension of this work to ω_0-accessible ordinals again requires no new idea.

Throughout this section individual variables range over ω_1. All lemmas are about the following <u>transition-system</u> (without initial and terminal condition). (Note the convention in Section 1, before Lemma 1.2.)

$$(1) \quad \Gamma_u^v(X,Z) \quad : \quad (\forall t)_u^v H[Xt,Zt,Zt'] \wedge (\forall x)_u^{v'} \mathcal{K}[\sup^x Z, Zx]$$

An $X[u,v)$-run of Γ is a sequence $Z[u,v]$ such that $\Gamma_u^v(X,Z)$; it is from a to b if $Zu=a$ and $Zv=b$; it is C-exact if $\{Zt; u\leq t<v\} = C$. For a given input X we are interested in the sets $S_u^{c,C}v$, of all states b which can be reached from c by a C-exact run. Thus, for $u\leq v$ we have,

(2) $S_u^{c,C}v$: $\{b; \exists Z[u,v].Zu=c \wedge \Gamma_u^v(X,Z) \wedge Zv=b \wedge \{Zt; u \leq t < v\} = C\}$

As in Section 3, we use the abbreviation $S_u v$ and we define the merging relations $u \approx v(t)$ and $u \approx v(-x)$. For the step from t to t' we still have the same recursion (4) Section 3. We state this as our first lemma.

Lemma 4.1: There are e, F (those in (4) Section 3) such that for any X and any $u \leq t$ the equations $S_u u=e$, $S_u t'=F[Xt, S_u t]$ hold.

Our first task is to add to this recursions for $S_u x$, in case x is a limit. Lemma 3.11 does this for $x=\omega$; it tells whether or not $b \in S_0 \omega$ in terms of values $S_u v$ before ω. Now, our proof of 3.11 easily extends from ω to any ω-limit x. One just has to replace throughout "finite subset of ω" by "not terminal subset of x", or correspondingly "infinite subset of ω" by "cofinal subset of x". I.e., do not confuse $(\exists^x t)$ with "there are infinitely many $t<x$"! There is just one point in the proof of 3.11, where the fact that x is ω-limit is used. Namely, in the second part of the proof of Lemma 3.4 we obtain partial runs $Z[y_i, y_{i+1}]$ for a x-cofinal $\{y_i\}$, and we desire to splice these to obtain a run $Z[u,x)$. Here we must be careful that no gap occurs. This is accomplished by choosing the y_i to be an ω-sequence approaching x. As we do not know how to use the second formula (9) of Section 3 for arbitrary x, we now outline a somewhat abbreviated proof of Lemma 4.3, our general version of Lemma 3.11.

<u>Remark</u> 4.2: Let X be arbitrary and let x be a limit. There is a $w < x$ such that $(\exists^x y) y \approx w(-x)$. In fact, there is an ω-sequence $y_0 < y_1 < y_2 < \cdots \to x$, such that $y_i \approx w(y_{i+1})$.

<u>Proof</u>: As in Remark 3.2 we show that $\approx (-x)$ is an equivalence of finite index. Therefore, some class must occur cofinal x. Let $w < x$ be a representative of such a class. So, $(\exists^x y) y \approx w(-x)$. As every limit $< \omega_1$ is an ω-limit, we therefore have an ω-sequence $v_0 < v_1 < v_2 < \cdots \to x$, such that $v_i \approx w(-x)$. So, to each i there is a $j > i$, such that $v_i \approx w(v_j)$. The sequence y_i, in the remark, can therefore be selected as a subsequence of the v_i's. Q.E.D.

We now define the following sets of states of Γ.

$$(3)\quad b\in B^{c,C}[s_0,s] \;:\; \bigvee_{d,D}[D\subseteq C \wedge \mathcal{X}[D,b] \wedge d\in s_0^{c,C}\cap s^{d,D}] \;.$$

For any input X, for any $u<x$, where x is ω-limit, we now show:

$$(4)\quad b\in S_u^{c,C}x \;.\!\equiv.\; (\exists w)^x(\exists^x y).y \approx w(-x) \wedge b\in B^{c,C}[S_u y, S_w y] \;.$$

<u>Proof</u>: Suppose first $b\in S_u^{c,C}x$. Then there is a run $Z[u,x]$ such that $Zu=c$, $\Gamma_u(X,Z)$, $Zx=b$, and $\{Zt;u\leq t<x\} = C$. By Remark 4.2 we have $w<y_0<y_1<\cdots\to x$, such that $y_i \approx w(-x)$, and even $y_i \approx w(y_{i+1})$. Now let $D=\sup^x Z$. Then clearly $D\subseteq C$, and as x is a limit and Z is a run of Γ ending in b, $\mathcal{X}[D,b]$. Note that $< S_u y_i, S_w y_i, Zy_i >$ takes on but finitely many values as $y_i\to x$. We may therefore assume that all these values are equal (else select a cofinal subsequence which has this property; it will preserve all earlier properties). Say $S_u y_i=s_0$, $S_w y_i=s$, $Zy_i=d$, for all y_i. As $D=\sup^x Z$ we may also assume $\{Zt;y_i\leq t<y_{i+1}\}=D$, for all i (else select cofinal subsequence). As $\Gamma_u^x[X,Z]$ this will imply $d\in S_{y_i}^{d,D}y_{i+1}$, and because $y_i \approx w(y_{i+1})$, $d\in S_w^{d,D}y_{i+1}$. Because $S_w y_{i+1}=s$, this yields $d\in s^{d,D}$. As $Z[u,x]$ is C-exact and $y_i\to x$, there is an i such that $C=\{t;u\leq t<y_i\}$. As $Zu=c$, Γ_u^x, $Zy_i=d$, this implies $d\in S_u^{c,C}y_i=s_0^{c,C}$. We now have, $D\subseteq C$, $\mathcal{X}[D,b]$, $d\in s_0^{c,C}\cap s^{d,D}$. Hence by (3), $b\in B^{c,C}[s_0,s]$. As $y_i\to x$, $y_i \approx w(-x)$, $S_u y_i=s_0$, $S_w y_i=s$, this implies the right side of (4).

Assume now the right side of (4). Using Remark 4.2 we thus have $w<y_0<y_1<y_2<\cdots\to x$ such that $y_i \approx w(y_{i+1})$ and

$b \in B^{c,C}[s_u y_i, s_w y_i]$. As x is an ω-limit we may assume that $\{y_i\}$ is an ω-sequence. We may also assume that $s_u y_i = s_0$, $s_w y_i = s$ take the same value for all i. Now $b \in B^{c,C}[s_0, s]$. Hence by (3), there are $D \subseteq C$, $\chi[D,b]$, $d \in s_0^{c,C}$, $d \in s^{d,D}$. As $d \in s_0^{c,C} = s_u^{c,C} y_0$, there is a C-exact run $Z[u,y_0]$ of Γ such that $Zu=c$, $Zy_0=d$. As $y_i \approx w(y_{i+1})$, we have $d \in s^{d,D} = s_w^{d,D} y_{i+1} = s_{y_i}^{d,D} y_{i+1}$. Hence, for every i, there is a D-exact run $Z[y_i, y_{i+1}]$ of Γ such that $Zy_i = Zy_{i+1} = d$. As these runs properly fit, and as $\{y_i\}$ is an ω-sequence approaching x, we can splice them to a run $Z[u,x)$. This clearly is a run of Γ, starting with $Zu=c$, and is C-exact (as $D \subseteq C$). It is also clear that $\sup^x Z = D$, and hence $\chi[\sup^x Z, b]$. Therefore, completing this run to $Z[u,x]$, by setting $Zx=b$, will still be a run of Γ. This run shows that $b \in s_u^{c,C} x$. Q.E.D.

The one place in this proof will be the only one where ω_0-accessibility is used. We thus have neatly isolated this matter from the very involved subset-construction. Note that of the partial run $Z[y_i, y_{i+1}]$ we only know existence. As it stands, the axiom of choice, namely $AC_{\omega_0}^{\gamma}$ where $\gamma = 2^{\omega_0}$, therefore is used. Formula (4) specializes to the first entry in (9), Section 3. The analog to the second entry in (9) is,

$$b \notin s_u^{c,C} x \; .\equiv. \; (\exists w)^x (\exists^x y).y \approx w(-x) \wedge b \notin B^{c,C}[s_u y, s_w y] \quad .$$

We do not need this. It is valid for ω-limits, and shows that $S_u x$ can also be expressed in the dual form to (4). This would have to come to use in a non-deterministic proof; does it exist?

We now are ready to transcribe our McNaughton proof in Section 3. For this purpose we need the exact merging relation $u \sim v(t)$, see formula (11) in Section 3. The analog to (12) is the following formula, which holds for ω-limits x, and $u<x$.

$$(5)\ b\in S_u^{c,C}x \ .\equiv.\ (\exists w)^x(\exists^x t)(\exists y)_u^x{}_{'}.y \sim w(t) \wedge b\in B^{c,C}[S_u y, S_w y]$$

Proof: (see proof of (12) in Section 3 for more details; we also drop the superscript c,C). Suppose first that $b\in S_u x$. By (4) we have a w and cofinal subset P of x such that, $Py\supset(\exists t)^x y \approx w(t)$ and $Py\supset b\in B(y) \wedge u<y$. We now define

$$y_i=(\mu y).Py \wedge (\forall j)^i t_j<y, \quad t_i=(\mu t)y_i \approx w(t).$$

This is just as in the proof of Remark 4.2. The induction halts at limit x, $y_i\to x$, Py_i, $y_i\leq t_i<y_{i+1}$, and clearly $y_i \sim w(t_i)$. Hence also $t_i\to x$, so we have

$$(\exists^x t)(\exists y)_u^x{}_{'}.y \sim w(t) \wedge b\in B(y_i).$$

This is (5) from left to right.

Suppose now the right side holds. So we have w and a cofinal subset T of x, and a function y_t from T into x, such that (c) $Tt\supset y_t \sim w(t)$, and (d) $Tt\supset b\in B(y_t) \wedge u<y_t$. From (c)

one proves that $A_z = \{t; Tt \wedge y_t \leq z \leq t\}$ is finite, for every $z<x$. Therefore $(\forall z)^x (\exists t)[Tt \wedge z<y_t]$. Hence by (c, d) $(\forall z)^x (\exists y)_z^x [y \approx w(-x) \wedge b \in B(y)]$. By (4) this implies $b \in S_u x$. Q.E.D.

We now introduce the piggyback device (13) in Section 3. The only difference is that we need many Q's.

$r_u t$: $(\mu z) z \approx u(-t)$ $\quad$ Ut : the sequence of all $r_u t$, $u<t$

(6) $u \leftarrow v(t)$: v in Ut, u the first in Ut such that $u \approx v(t)$

$Q^{c,C}_{u,w} t$: $\{ < b, r_y t >; u, w \leq y < t \wedge b \in B^{c,C}[S_u y, S_w y]\}$

In the following two proofs we will use the basic properties (i), (ii), etc. listed after (13) in Section 3. They are easily extended to transfinite ordinals. We now show the new form of $S_u x$, for ω-limits x, and $u<x$.

(7) $b \in S^{c,C}_u x \ .\!\equiv.\ (\exists w)^x . (\forall^x t) w \leftarrow w(t) \wedge (\exists^x t)(\exists v)[v \neq w \leftarrow v(t) \wedge \langle b, v \rangle \in Q^{c,C}_{u,w} t]$

<u>Proof</u>: (We drop the superscript c,C). Assume first that $b \in S_u x$. By (5) there are $w<x$, cofinal x many t_i, and $y_i<x$, such that (a) $y_i \sim w(t_i)$ and $b \in B(y_i)$. We may assume $w<t_i$, and $w=r_w x$ is the first in its class modulo $\approx (-x)$. By (iii) we have (b), $(\forall t)^x_w, w \leftarrow w(t)$, and in particular $w \leftarrow w(t_i)$. By (jw), $t_i \not\approx w(t_i)$ and by (a), $y_i < t_i$. Therefore $v_i = r_{y_i} t_i$ exists, and by (i), $v_i \leq y_i < t_i$ and $y_i \approx v_i(-t_i)$. By (a) this implies $v_i \neq w$ and $v_i \approx w(t_i)$. Hence by (ii) and $w \leftarrow w(t_i)$, we have $w \leftarrow v_i(t_i)$. So by (6), $w \leq v_i$, and we have $w \leq y_i < t_i$. As $b \in B(y_i)$ we use

(6) to see $< b, v_i > \in Q_{u,w}t_i$. Because of (b), and as the t_i are cofinal x this yields the right side of (7).

Assume now the right side of (7). So we have $w<x$, cofinal x many t_i, and $v_i<x$ such that $w \vdash w(t_i)$, $v_i \neq w$, $w \vdash v_i(t_i)$, and $< b, v_i > \in Q_{u,w}t_i$. By (6) there exists y_i such that $v_i = r_{y_i}t_i$, $w \leq y_i < t_i$, and $b \in B(y_i)$. Hence, $v_i \approx y_i(t_i)$ and $w \approx v_i(t_i)$, so $y_i \approx w(t_i)$. As $w \vdash w(t_i)$ and $w<t_i$ we use (j) to get $r_w t_i = w \neq v_i = r_{y_i} t_i$. Hence by (iv), $y_i \not\approx w(-t_i)$. So $y_i \sim w(t_i)$. As $b \in B(y_i)$, and the t_i are cofinal x, we use (5) to conclude $b \in S_u x$. Q.E.D.

$S_u x$ is now essentially in the right form. It remains to supply recursions for the piggyback components. In particular we now also need such at limits.

<u>Lemma</u> 4.3: The sequences U, Q defined by (6) satisfy the following recursions (x an ω-limit):

$U0$ = empty sequence $\qquad Q^{c,C}_{u,w}w = 0$

Ut' = remove from Ut all $u \nvdash u(t)$, and add t at the end.

Ux = all u such that $(\forall^x t) u \vdash u(t)$, in natural order.

$$Q^{c,C}_{u,w}t' = \{< b, v >; (\exists p)[v \vdash p(t) \wedge < b, p > \in Q^{c,C}_{u,w}t] \vee [v=t \wedge b \in B^{c,C}[S_u t, S_w t]]\}$$

$$Q^{c,C}_{u,w}x = \{< b, v >; (\forall^x t) < b, v > \in Q^{c,C}_{u,w}t\}$$

<u>Proof</u>: The initial equations are trivial, by (6). The equations at t' are proved just as in Lemma 3.12. We show the proofs for

the equations at ω-limits x, using the items (i), (ii), etc. stated after (13) in Section 3.

Suppose $u \in Ux$. By (6) there is a $y<x$ such that $u=r_y x$. By (iii), $(\forall t)^x_w, u \leftarrow u(t)$. Hence $(\forall^x t) u \leftarrow u(t)$. Suppose now $(\forall^x t) u \leftarrow u(t)$. Let $v=r_u x$. By (i), $v \leq u$ and $v \approx u(-x)$. As x is a limit we therefore have a $t<x$ such that $u \leftarrow u(t)$ and $v \approx u(-t)$. Hence by (iv) and (j), $u=r_u t=r_v t$. Therefore by (i) $u \leq v$. So $u=v=r_u x$, and therefore $u \in Ux$. This argument proves the equation for Ux. It remains to show that for $Q^{c,C}_{u,w} x$. (We drop the superscript c,C.)

Suppose $< b,v > \in Q_{u,w} x$. By (6) there is a y such that $v=r_y x$, $u,w \leq y<x$, and $b \in B(y)$. So $v \approx y(-x)$, and as x is limit there is a $z<x$ such that $(\forall t)_z v \approx y(-t)$. Therefore by (iv), $(\forall^x t) r_v t=r_y t$. As $v=r_y x$ we have $v \in Ux$, and we have just seen that this yields $(\forall^x t) v \leftarrow v(t)$. So by (j), $(\forall^x t) v=r_v t$. We conclude, $(\forall^x t) v=r_y t$. As $u,w \leq y<x$ and $B(y)$ this yields by (6), $(\forall^x t) < b,v > \in Q_{u,w} x$. This argument shows the equation for $Q_{u,w} x$ holds in one direction. To show the other direction, assume $(\forall^x t) < b,v > \in Q_{u,w} t$. From (6) it clearly follows $(\forall^x t)(\exists y) v=r_y t$. Hence $(\forall^x t) v \in Ut$, and by the equation for Ux we have just shown, $v \in Ux$. From our assumption and (6) we also have a t and y such that $u,w \leq y<t<x$, $v=r_y t$, and $B(y)$. As $v \in Ux$ we have by (vi), $v=r_y x$. $b \in Q_{u,w} x$ now follows by (6). Q.E.D.

We are now ready to set up a finite state recursion for ω_1-sequences S_i, V, $Q_{i,k}$ which simulate the sequences S_u, U, $Q_{u,w}$.

$$S_i t = \begin{cases} e & \text{if } i \text{ not in } Vt \\ \mathcal{S}_i(t) & \text{if } i \text{ in } Vt \end{cases}$$

$$V0 = \vec{0}$$

(8) $Vt' =$ remove from Vt all $i \nless i(t)$;
add $\zeta(t)$ at end

$Vx =$ all i such that $(\forall^x t)\, i < i(t)$;
i before j in case $(\forall^x t)[i \text{ before } j]$

$$Q_{i,k} t = \begin{cases} \overline{0} & \text{if } i \text{ or } k \text{ not in } Vt \\ \mathcal{Q}_{i,k}(t) & \text{if } i,k \text{ in } Vt \end{cases}$$

Here the indices i, j, k, h range over $\{0,\ldots,g\}$, $e^{c,0} = \{c\}$ and $e^{c,C} = 0$ if $C \neq 0$, $\vec{0}$ is the empty vector, $\overline{0} = \langle 0,0,\ldots,0 \rangle$, and the following abbreviations are used:

$$
\begin{aligned}
S_i t &: \langle S_i^{c,C} t; c \text{ a state of } \Gamma,\ C \text{ a set of states of } \Gamma \rangle \\
Q_{i,k} t &: \langle Q_{i,k}^{c,C} t; c \text{ a state of } \Gamma,\ C \text{ a set of states of } \Gamma \rangle \\
i \leftarrow j(t) &: j \text{ in } Vt, \text{ and } i \text{ is the first in } Vt \text{ such that } S_i t = S_j t \\
\zeta(t) &: \text{the first } i \leq g, \text{ such that } i \text{ not in } Vt \\
\mathcal{S}_i^{c,C}(t') &: \{b; \bigvee_a .a \in C \wedge H[Xt,a,b] \wedge a \in (S_i^{c,C} t \cup S_i^{c,C-a} t)\} \\
\mathcal{S}_i^{c,C}(x) &: \{b; \bigvee_k .(\forall^x t) k \leftarrow k(t) \wedge (\exists^x t) \bigvee_j [j \neq k \leftarrow j(t) \wedge \langle b,j \rangle \in Q_{i,k}^{c,C} t]\} \\
\mathcal{Q}_{i,k}^{c,C}(t') &: \{\langle b,j \rangle; \bigvee_h [j \leftarrow h(t) \wedge \langle b,h \rangle \in Q_{i,k}^{c,C} t] \vee \\
&\qquad [j = \zeta(t) \wedge b \in B[S_i t, S_k t]]\} \\
\mathcal{Q}_{i,k}^{c,C}(x) &: \{\langle b,j \rangle; (\forall^x t) \langle b,j \rangle \in Q_{i,k}^{c,C} t\}
\end{aligned}
$$

As each component S_i, V, $Q_{i,j}$ takes on states in a prescribed finite set, and by the remark on sup-conditions, Lemma 1.2, the recursion (8) clearly is of the following form.

$$
(9) \qquad Y0 = a, \quad Yt' = F[Xt, Yt], \quad Yx = \mathcal{S}[\sup^x Y]
$$

Here Y takes on states in a prescribed finite set, F maps pairs of input-states and states into states, and $\mathcal{S}$ maps sets of states to states. We will call these <u>finite-state recursions</u>.

<u>Lemma</u> 4.4(<u>the subset-construction for countable ordinals</u>). To any transition-system Γ of form (1) with initial condition $E[\cdot]$ and terminal condition $L[\cdot]$, one can construct a finite-state recursion A of form (9) with terminal condition $L'[\cdot]$ such that, for any countable ordinal α and any input $X[0,\alpha)$, $[E,\Gamma,L]$ accepts X if and only if $[A,L']$ accepts X. That is

$(\exists Z).E[Z0] \wedge \Gamma_0^{\alpha}(X,Z) \wedge L[Z\alpha]$ holds just in case the recursion $\mathcal{A}$ applied to X yields a final state $Y\alpha$ which satisfies $L'[Y\alpha]$. In fact, the recursion (8) is such an $\mathcal{A}$, the terminal condition L' being

$$\bigvee_{c,C,b} .E[c] \wedge b \in L \cap S_0^{c,C}\alpha \quad .$$

<u>Proof</u>: We ask the reader to make up the formulas $(\overline{8})$, obtained from (8) by making the following replacements: i,j,k,h by u,v,w,p; $S,\mathcal{J},Q,\mathcal{Q}$ by $\overline{S},\overline{\mathcal{J}},\overline{Q},\overline{\mathcal{Q}}$; V by U; $\leftarrow$ by $\Leftarrow$; $\zeta(t)$ by t. Now let $\alpha<\omega_1$ and let $X[0,\alpha)$ be an arbitrary α-sequence. Let $\overline{S}_u, U, \overline{Q}_{u,w}$ be the α'-sequences defined by the recursion $(\overline{8})$, and let S_i, V, $Q_{i,k}$ be the α'-sequences defined by the recursion (8).

Note that $(\overline{8})$ contains the recursions which occur in Lemmas 4.2 and 4.3, and in (7). By the results stated in these Lemmas and (7) it therefore follows that the sets $\overline{S}_0^{c,C}\alpha$ are just those introduced in (2) (note that $0 \in U\alpha$). So we have,

(a) $\quad b \in \overline{S}_0^{c,C}\alpha \quad .\equiv. \quad (\exists Z).Z0=c \wedge \Gamma_0^{\alpha}(X,Z) \wedge Z\alpha=b$.

The recursion $(\overline{8})$ together with the terminal condition $\bigvee_{c,C} [E[c] \wedge b \in L \cap \overline{S}_0^{c,C}\alpha]$ would therefore answer the requirement in our lemma, except $(\overline{8})$ is not finite state. The lemma will be proved if we can show $\overline{S}_0\alpha = S_0\alpha$. To see this we introduce the function $u_i t$:

(c)
$$i=\zeta(t) .\supset. u_i t' = t \qquad i \leftarrow i(t) .\supset. u_i t' = u_i t$$
$$(\forall^x t) i \leftarrow i(t) .\supset. u_i x = \text{that } u, \quad (\forall^x t) u_i t = u$$

This function shows in just which way the finite-state recursion

(8) simulates the "quasi finite-state" recursion $(\overline{8})$. Namely, we will prove:

(b_1) i in $Vt . \supset . u_i t$ defined, in Ut

(b_2) u in $Ut . \supset . (\exists i)[i$ in $Vt \wedge u_i t = u]$

(b_3) i before j in $Vt . \supset . u_i t < u_j t$

(b_4) i in $Vt \wedge u = u_i t . \supset . S_i t = \overline{S}_u t$

(b_5) j in $Vt . \supset . i \leftarrow j(t) \equiv u_i t <= u_j t(t)$

(b_6) i,k in $Vt \wedge u = u_i t \wedge w = u_k t . \supset . < b, j > \in Q^{c,C}_{i,k} \equiv < b, u_j t > \in \overline{Q}^{c,C}_{u,w} t$

At $t=0$ these formulas are trivial, as both $U0$ and $V0$ are the empty vector. The inductive steps, from t to t', are just as in the case $\alpha=\omega$, and work exactly as shown in Lemma 3.13. We therefore present here the inductive step to a limit $x \leq \alpha$.

(b_1): Suppose i is in Vx. Then by (8), there is a $z<x$, such that $z \leq t < x \supset i \leftarrow i(t)$. Hence i is in Vz, and by inductive assumption $u = u_i z$ is defined in Uz. By (c) we have $z \leq t \leq x \supset u_i t = u$, and by inductive assumption $z \leq t < x \supset u_i t <= u_i t(t)$. Hence by $(\overline{8})$, $u \in Ux$. So $u_i x$ is defined and in Ux.

(b_2): Suppose u is in Ux. Then by $(\overline{8})$, there is a $z<x$, such that $z \leq t < x \supset u <= u(t)$. Hence u is in Uz, and by inductive assumption there is an i in Vz such that $u = u_i z$. Using these facts, the inductive assumption, and (c) one shows by simultaneous induction on t that $z \leq t \leq x . \supset . u_i t = u$, and $z \leq t < x \supset i \leftarrow i(t)$. Hence, by (8), i is in Vx. As $u_i x = u$ this settles b_2.

(b_3): Suppose i occurs before j in Vx. By (8) we therefore have a $z<x$ such that $z \leq t < x \supset . i \leftarrow i(t) \wedge j \leftarrow j(t) \wedge i$ before j in Vt. The argument for (b_1) shows that $u_i x = u_i z$ and $u_j x = u_j z$, and by inductive assumption we have $u_i z < u_j z$. Hence $u_i x < u_j x$.

(b_4): Suppose i is in Vx and $u = u_i x$. As we have shown in the argument for (b_1) it follows that u is in Ux, and $u_i t = u$ holds x-terminally. Similarly, if $(\forall^x t) u_k t \leftarrow u_k t(t)$ then there is a w in Ux such that $u_k t = w$ holds x-terminally. By supposition and (8), $S_i x$ is $\mathcal{J}_i(x)$, and by the remarks we therefore have,

$$b \in S_i x \;.\equiv.\; (\exists w) . \forall^x t) w = w(t) \wedge (\exists^x t) \bigvee_j [u_j t \neq w = u_j t(t) \wedge < b, u_j t > \in \overline{Q}^{c,C}_{u,w} t]$$

As u is in Ux, $\overline{S}_u x$ is $\overline{\mathcal{J}}_u x$. Hence the expression δ on the right side implies (using the inductive assumptions) $b \in \overline{S}_u x$. To see $b \in \overline{S}_u x \supset \delta$ we need in addition the fact (*) $< b, v > \in \overline{Q}^{c,C}_{u,w} t \supset v \in Ut$ (which is easily shown by induction on t). From the $v \in Ut$ we then obtain (by inductive assumption) j such that $v = u_j t$. This indicates the proof of $S_i x = \overline{S}_u x$.

(b_5): This is a consequence of the other parts of (b); see proof of Lemma 3.13.

(b_6): Suppose i,k are in Vx, $u = u_i x$, and $w = u_k x$. By the argument for (b_1) we know that, i in Vt, j in Vt, $u = u_i t$, $w = u_k t$ hold x-terminally. Hence, by inductive assumption, $< b, j > \in Q^{c,C}_{i,k} t \equiv < b, u_j t > \in \overline{Q}^{c,C}_{u,w} t$ holds terminally. Using (8),

$(\overline{8})$ and the remark (*) this yields the required equivalence at $t=x$.

We now have shown the formulas (b) for all $t \leq \alpha$. It remains to note that 0 is in all Vt, from $t=1$ on, and $u_0 t=0$. By (b_4) this yields the required $S_0\alpha=\overline{S}_0\alpha$. Q.E.D.

It is important for our work on ω_1, in Section 6, that the subset-construction work uniformly for all $\alpha<\omega_1$. This also yields the following strong form of the definability theorem for MT of countable ordinals.

Theorem 4.5 (definability in countable α): To every MT-formula $\Sigma(X)$ in the primitive $<$, one can construct a finite automaton $\mathfrak{A} = [a,F,\mathfrak{F},W]$, consisting of a finite-state recursion (9), and a terminal set W (of states), such that for any countable ordinal α and any $X[0,\alpha)$, $\Sigma(X)$ holds in $[\alpha,<]$ just in case $\mathfrak{A}$ accepts the input $X[0,\alpha)$; that is the recursion (9) yields a $Y\alpha$ which belongs to W.

Proof: We first use the prenex form Lemma 1.5 to put $\Sigma(X)$ into the form

$$(\text{predicate prefix } Y)(\exists Z).A[Z0] \wedge \Gamma_0^{\alpha}(X,Y,Z) \wedge D[Z\alpha].$$

By the remarks in Section 2, our Lemma 4.3 yields a complementation lemma, which works uniformly for all $\alpha<\omega_1$. The prefix Y can now be eliminated stepwise, as remarked in Section 2. Using Lemma 4.3 again, we obtain the required automaton $\mathfrak{A}$. Q.E.D.

It is important to note that the converse to Theorem 4.5 also holds: To any automaton $\mathcal{A}$ one easily makes up an MT-formula $\Sigma(X)$ in the primitives $<$, 0, ', Lm such that, for any ordinal α and any $X[0,\alpha)$, $\Sigma(X)$ holds in $[\alpha,<]$ just in case $\mathcal{A}$ accepts X. The primitives 0, ', Lm may be eliminated by using their definition from $<$.

Define the ω_1-behavior $beh_{\omega_1}\mathcal{A}$ of an automaton to be the set consisting of all inputs $X[0,\alpha)$, accepted by $\mathcal{A}$. These behaviors are the analogs of the regular events studied in automata theory. The extension of this theory to ω_1-events, consisting of well-ordered strings of lengths $<\omega_1$, now is in order. Our Lemma 4.4 is a fundamental result in this theory: An ω_1-event which is the behavior of a non-deterministic system, is also the behavior of an automaton. Theorem 4.5 says that questions about definability in the MT of countable ordinals are questions about ω_1-regular events.

We will now analyze what Theorem 4.5 does in the special case of MT-sentences Σ. That is, we will study the ω_1-behavior of input-free automata. Note that "input-free" means "just one input-state". So an input simply is an ordinal α, and behaviors are sets of ordinals. Define the ω_1-spectrum of an MT-sentence Σ (in the primitive $<$) to be the set of all ordinals α such that Σ holds in $[\alpha,<]$. By Theorem 4.5 we have:

Remark 4.6: To every MT-sentence Σ in $<$, one can construct an input-free automaton $\mathbf{A}$, such that the ω_1-spectrum of Σ is equal to the ω_1-behavior $beh_{\omega_1}\mathbf{A}$.

An input-free automaton consists of an input-free finite-state recursion of the following form,

$$(10) \qquad Z0=a, \quad Zt'=F[Zt], \quad Zx=\mathcal{G}[\sup^x Z]$$

and a terminal set W of states of Z. An ordinal α is accepted by (is in the behavior of) this automaton just in case $W[Z\alpha]$. We will now find a simpler form of the recursion (10), which shows more clearly how the run Z looks.

In the sequel K is a finite set, F is an operator from K to K, and $\mathcal{G}$ is an operator from subsets of K to K. For any $c \in K$ let Z_c be the sequence defined by (10), with the initial value a replaced by c. We now define the operators F_ν on K, for any ordinal ν:

$$(11) \qquad F_\nu[c]=Z_c\omega^\nu$$

That is $F_\nu[c]$ is the goal of the unique ω^ν-run from c. Note that $F_0=F$. These operators satisfy the following <u>absorption laws</u>:

$$(12) \qquad \nu<\mu \ .\supset. \ F_\mu F_\nu=F_\mu$$

Just note that, because $\nu<\mu$, $Z_c[\omega^\nu,\omega^\mu]$ is an ω^μ-run of $[F,\mathcal{G}]$ starting from $d=Z_c\omega^\nu$. As there is exactly one such, namely

$Z_d[0,\omega^\mu]$, we have $Z_c[\omega^\nu,\omega^\mu] = Z_d[0,\omega^\mu]$. Hence $Z_c\omega^\mu = Z_d\omega^\mu$. So $F_\mu[c] = F_\mu[d] = F_\mu F_\nu[c]$. As c was arbitrary this proves (12). We now show that the F_ν's become equal from some $p<\omega$ on.

<u>Lemma</u> 4.7: There is a finite p such that $F_p^2 = F_p$ and for every $c\in K$, $F_p[c] = \mathscr{S}[D_c]$ whereby $D_c = \{Z_c t; \omega^p \leq t < \omega^p 2\}$. Furthermore, for any such p we have $F_\nu = F_p$, for all $\nu \geq p$.

<u>Proof</u>: Let $Zt = \langle Z_c t; c\in K\rangle$ and let $D = \sup^z Z$ at $z=\omega^\omega$. Then clearly there are $u, v<\omega^\omega$, such that $v\neq 0$, and $Zu=Z(u+v)$, and $\{Zt; u\leq t<u+v\} = D$, and $u\leq t \supset Zt\in D$. It follows that $Z[u+vi, u+vi+v] = Z[u,u+v]$, for all $i<\omega$. Hence the set D keeps reoccurring as values of Z before $w=u+v\omega$. So $D=\sup^w Z$. As $u\leq w$ we have $Zw\in D$, and therefore the state Zw must occur between u and $u+v$. So, there is an x, such that $Zw=Zx$ and $u\leq x<u+v$. It follows that $\{Zt; x\leq t<w\} = D$. Let D_c be projection of D onto the c-axis. Then clearly $\{Z_c t; x\leq t<w\} = D_c$, and $D_c = \sup^w Z_c$ (as $D = \sup^w Z$). Hence, $Z_c w = \mathscr{S}[D_c]$. Letting y be such that $w=x+y$, we thus have,
(a) $x,y<\omega^\omega$, y a limit, $Z_c x = Z_c(x+y) = \mathscr{S}[D_c]$, $D_c = \{Z_c t; x\leq t<x+y\}$.
From this we obviously have $Z_c(x+yi) = Z_c x$, for all i. And, as the value sets D_c keep repeating, $\sup^{x+y\omega} Z_c = D_c$, and hence $Z_c(x+y\omega) = \mathscr{S}[D_c] = Z_c x$. This can be extended by induction. So (a) remains true if y is replaced by any $\bar{y}=y\omega^i$, and x is replaced by $\bar{x}=x+\bar{y}$. As $x<\omega^\omega$ we can choose $i<\omega$ such that

$x<\bar{y}$, and as $y<\omega^{\omega}$, $\bar{y}$ will be of form ω^p if i is chosen $\neq 0$. Now we have $\bar{y}=\omega^p$, and as $x<\bar{y}$, $\bar{x}=x+\omega^p=\omega^p$. So we have a $p<\omega$ such that for all c, $Z_c\omega^p = Z_c(\omega^p+\omega^p)$ and $\{Z_c t; \omega^p \leq t < \omega^p 2\} = D_c$ and $Z_c\omega^p = \mathcal{J}[D_c]$. By the definition of F_p, the first part can be restated as $F_p^2 = F_p$, and the last part as $F_p[c] = \mathcal{J}[D_c]$. This proves the first part of our lemma.

Suppose now that p has the property in the first part of the lemma. Then (a) holds for $x=y=\omega^p$. As we have already noted, (a) therefore also holds for $x=\omega^p$ and $y=\omega^p\cdot\omega^\nu$, ν arbitrary. Hence, for all ν and c, $Z_c\omega^p = Z_c\omega^{p+\nu}$, i.e., $F_p[c] = F_{p+\nu}[c]$. Thus, $F_p = F_{p+\nu}$ for all ordinals ν. Q.E.D.

__Theorem__ 4.8 (__the normal form of sentences__): Let Σ be any MT-sentence in the primitive $<$. There exist a finite number p, and operators $F_0,\ldots,F_p$ on a finite set K, $a\in K$ and $W\subseteq K$, such that for any countable ordinal $\alpha=\omega^p\nu+\omega^{p-1}k_{p-1}+\cdots+\omega^0 k_0$, Σ holds in $[\alpha,<]$ just in case $F_0^{k_0}\ldots F_{p-1}^{k_{p-1}}F_p^{(\nu)}[a]\in W$. Here (ν) stands for 0 or 1, according to whether $\nu=0$ or $\nu>0$. Furthermore, the operators F_i may be made to satisfy the absorption laws (12), and $F_p^2 = F_p$.

__Proof__: By Remark 4.6 we can find a recursion (10) with a finite set K of states, and a $W\subseteq K$, such that $[a,F,\mathcal{J},W]$ accepts just those α in the ω_1-spectrum of Σ. Let Z be the unique run of (10). So we have for every countable α, Σ holds in $[\alpha,<]$ just in case $Z\alpha\in W$. Now, let F_ν be defined by (11). So (12) holds, and it is obvious that, for any ordinal ν, Z satisfies

$Z(t+\omega^{\nu}) = F_{\nu}[Zt]$. By Lemma 4.7, there is a $p<\omega$ such that $F_{\nu} = F_p$ for $\nu \geq p$. Hence, by (12) $F_p^2 = F_p$, and Z satisfies the following formulas:

$$(10') \quad Z0=a, \quad Z(t+\omega^0)=F_0[Zt],\ldots,Z(t+\omega^{p-1})=F_{p-1}[Zt], \; Z(t+\omega^{p+\zeta})=F_p[Zt]$$

Suppose now that $\nu<\omega_1$ and $\alpha=\omega^p\nu+\omega^{p-1}k_{p-1}+\ldots+\omega^0k_0$. Let $\nu=\omega^{p_s}h_s+\ldots+\omega^{p_0}h_0$ be the ω-expansion of ν. From (10') we obtain $Z\alpha=F_0^{k_0} \ldots F_{p-1}^{k_{p-1}}F_p^{(\nu)}[a]$, whereby $(\nu)=h_s+\ldots+h_0$. So $(\nu)=0$ if $\nu=0$, and as $F_p^2 = F_p$, (ν) can be replaced by 1 if $\nu\neq 0$. As Σ holds in $[\alpha,<]$ just in case $Z\alpha\in W$, this proves our theorem. Q.E.D.

Note that for any p, every ordinal α has a unique representation $\alpha=\omega^p\nu+\omega^{p-1}k_{p-1}+\ldots+\omega^0k_0$, we call this the p-<u>expansion</u> of α. $\omega^p\nu$ is the p-<u>head of</u> α, and the rest of the expansion is the p-<u>tail of</u> α. The sequence $<(\nu),k_{p-1},\ldots,k_0>$ is called the p-<u>character</u> of α. We also have a unique representation $\alpha=\omega^{\omega}\nu+\omega^{q-1}k_{q-1}+\ldots+\omega^0k_0$, whereby $q=0 \vee k_{q-1}\neq 0$. $\omega^{\omega}\nu$ is the ω-<u>head</u>, the rest is the ω-<u>tail</u> of α, and $<(\nu),k_{q-1},\ldots,k_0>$ is the ω-<u>character</u> of α.

Consider now the following absorption laws:

$$(12') \qquad i<j\leq p.\supset.F_jF_i=F_j \qquad F_pF_p=F_p$$

These are just the requirements (unique solvability) which make (10') a proper recursive definition of a sequence Z of states.

As shown in the proof of 4.8, every input-free finite-state recursion (10) can be replaced by such a <u>normal recursion</u> (10'), the translation being given by (11); the existence of p was shown in Lemma 4.7. Conversely, it is not difficult to translate (10') into form (10), using the absorption laws (12'). (For example, state $(\exists Z)10'$ as an MT-sentence and use Theorem 4.8.) Now the normal form (10') is really much easier to understand (for example it yields the nice information in 4.8). The system $[K,a,F_0,\ldots,F_p]$ is a finite transition-algebra working on words over the input states $\{0,\ldots,p\}$ in the usual sense of finite automata theory. If you like, you may replace the totally free algebra with $(p+1)$ successor functions, by that which is free relative to the absorption laws (12'). In this case an input, instead of being a word over $\{0,\ldots,p\}$, becomes a p-character $<\zeta,k_{p-1},\ldots,k_0>$, ζ equals 0 or 1 and k_i any natural number. Call these automata $[K,a,F_0,\ldots,F_p,W]$ absorption-automata. Their behaviors are sets of p-characters, and Theorem 4.8 takes the following appealing combinatorial form.

<u>Theorem</u> 4.8': The ω_1-spectrum of any MT-sentence Σ consists of all $\alpha<\omega_1$ whose p-characters are accepted by an appropriately chosen p-absorption automaton.

To every MT-sentence Σ we can associate the smallest number $p(\Sigma)$ which works for Σ as in 4.8. This is obviously an important invariant. For example, from 4.8 we see: If Σ has a countable model $[\alpha,<]$, then it must have one with $\alpha<\omega^{p+1}$,

$p=p(\Sigma)$. So the structures $[\alpha,<]$ with $\alpha<\omega^\omega$, are dense in the MT-space of all countable well-orders. I.e., the MT of all countable well-orders is equal to MT of all $\alpha<\omega^\omega$.

Let $\Sigma[x]$ be the relativization of the sentence Σ to x. The other way around, given a formula $\Phi(x)$ it is easy to make up a sentence Σ such that $\Phi(x)$ is equivalent to $\Sigma[x]$. For any α, the α-spectrum of Σ is just the set defined by $\Sigma[x]$ in $[\alpha,<]$. Hence, in the case of countable ordinals, discussing spectra of sentences comes to discussing definability by formulas with one free individual variable. For example we can restate 4.8' as:

Theorem 4.8": The following four Boolean algebras are isomorphic: the algebra of ω_1-spectra of MT-sentences, the Tarski-Lindenbaum algebra of the MT-theory of all countable well-orders, the algebra of all sets of ordinals MT-definable in $[\omega^\omega,<]$, the algebra of all absorption-regular sets.

The last of these consists of all behaviors of absorption-automata (p arbitrary), and is by far the most concrete presentation of the Boolean algebra considered. In a joint paper with D. Siefkes, refered to as [BS], we will present a careful study of this algebra. In particular, all maximal ideals will be described. These are just the complete elementary extensions of the theory MT(co) of all countable well-orders. Most of these are found to admit no standard model.

From [4] one can see that precisely the same Boolean algebra comes up in WMT, in the same manner as in 4.8". Only, now it is also the algebra of the WMT of all ordinals. This was observed by D. Siefkes. From 4.8 and [4] one thus obtains the following information.

Theorem 4.9: For any countable ordinals α and β, the following statements are equivalent.

1. The systems $[\alpha,<]$ and $[\beta,<]$ are MT-equivalent.
2. The two systems are WMT-equivalent.
3. α and β have the same ω-character.
4. Either $\alpha=\beta<\omega^{\omega}$ or else $\omega^{\omega}\leq\alpha,\beta$ and α,β have the same ω-tail.

In the WMT-case this extends to all ordinals. In particular ω^{ω} is WMT-equivalent to ω_1 and to $\varkappa_1$, the first ordinal which is not accessible from 0 by ' and ω_0-limits. As pointed out in Section 1 this situation changes drastically: $[\varkappa_1,<]$ is MT-categorical and finitely axiomatizable. Note $Ac_{\omega_0}^{\omega_0}$ implies $\varkappa_1 = \omega_1$. A. Litman has shown that, even without AC, $[\omega_1,<]$ is not MT-equivalent to $[\alpha,<]$ for any countable α. Going over our proofs, we find just one place where ω_0-accessibility was used; namely in the splicing process in the proof of (4). In this same connection we used AC, and nowhere in the proof do we need to refer to the cardinality of ordinals. At one time we thought that AC is not really needed for making the selection of partial runs to be spliced. From this it would follow that all the results in this section go through without using AC, and that all results hold in the stronger form, in which ω_1 is replaced by $\varkappa_1$, i.e., "countable ordinal" is replaced by "ω_0-accessible ordinal". If $\omega_1 < \varkappa_1$ was added as an axiom we would thus have: $MT[\omega_1,<] = MT[\omega^\omega,<]$, which contradicts Litman's result. Hence Litman has shown that some part of AC must be used in the splicing process in (4).

Just how little more set theory is needed to make our results go through will be shown in [BS], by presenting a simple elementary axiom system for MT(co). The task is to comb through the proof, trying to formalize as an MT-sentence each property of ordinals used.

Item 3: Decision methods: There is just one place in our long proof of Theorem 4.8, where we failed to indicate an actual construction from Σ of the items leading to the absorption automaton $[K,a,F_0,\dots,F_p,W]$. Namely, the proof of Lemma 4.7 does not tell how p and $F_0,\dots,F_p$ may be constructed from $[F,\mathfrak{s}]$. We now will show how the finite objects F_0, F_1, $F_2,\dots$, defined by (11), can be actually obtained, one after the other. In each case we will show how to construct the sets

$$D_{c,i} = \{Z_c t; \omega^i \leq t < \omega^i 2\} .$$

According to the second part of Lemma 4.7, we may stop the process at the first p, such that $F_p^2 = F_p$ and $F_p[c] \doteq \mathcal{S}[D_{c,i}]$ for each c. The first part of Lemma 4.7 tells that such a stop will occur.

Lemma 4.10: Given a finite set K, an operator $F:K\to K$, an operator $\mathcal{S}:\mathcal{O}K\to K$. For each $c\in K$ let z_c be defined by $z_c 0=c$, $z_c t'=F[z_c t]$, $z_c x=\mathcal{S}[\sup^x z_c]$. Let F_i, V_i, S_{i+1} be the following operators. $F_i[c]=z_c\omega^i$. $V_i[c]=\{z_c t; 0\leq t<\omega^i\}$. $S_{i+1}[c]=\sup^{\omega^{i+1}} z_c$. These operators can be constructed by the following simultaneous recursion:

$$F_0[c] = F[c] \qquad\qquad V_0[c] = \{c\}$$

$$<h_i,k_i> = \text{the first pair } h<k, \text{ such that for all } c\in K,\ F_i^h[c]=F_i^k[c]$$

$$S_{i+1}[c] = V_i[F_i^{h_i}[c]]\cup\cdots\cup V_i[F_i^{k_i-1}[c]]$$

$$V_{i+1}[c] = V_i[F_i^0[c]]\cup\cdots\cup V_i[F_i^{h_i-1}[c]] \cup S_{i+1}[c]$$

$$F_{i+1}[c] = \mathcal{S}[S_{i+1}[c]]$$

Proof: The initial formulas obviously hold. Because K is finite the pair $<h_i,k_i>$ does exist, for each F_i. Note that $F_i^h[c]$ is the same as $z_c(\omega^i h)$. So all z_c repeat their value at $\omega^i h_i$, $\omega^i k_i$. Hence z_c becomes periodic with the period $z[\omega^i h_i,\omega^i k_i)$. Therefore $\{z_c t; \omega^i h_i\leq t<\omega^i k_i\}=S_{i+1}[c]$. From this the equations for $S_{i+1}[c]$ and $F_{i+1}[c]$ clearly follow. $V_{i+1}[c]$ is obtained by adding $\{z_c t; 0\leq t<\omega^i h_i\}$ to the sup $S_{i+1}[c]$. That is what the equation for $V_{i+1}[c]$ says. Q.E.D.

Note that this was just a repetition of the proof of (4), in the special case where no input X is present. Lemma 4.11 together with the stop-information in Lemma 4.7 show how one can effectively construct the absorption-automaton (10') from the automaton (10).

<u>Theorem</u> 4.11 (<u>The decision method</u>): There is a method which applies to any MT-sentence Σ, and to any ω-character π, and which decides whether or not Σ holds in the system $[\alpha,<]$, α a countable ordinal of ω-character π.

<u>Proof</u>: Going over our proofs, which lead to Remark 4.6, one can make up a procedure which constructs the input-free automaton $A = [a, F, \mathcal{S}, W]$ from Σ. We now use the process in Lemma 4.10 to construct $F_0, V_0, S_1, V_1, F_1, S_2, V_2, F_2, \cdots\cdots$. This process is stopped when we arrive at a p, such that $F_p^2 = F_p$ and, for all c, $F_p[c] = \mathcal{S}[V_p F_p[c]]$. By Lemma 4.11, we know such a p must come up, and that $a, F_0, \ldots, F_p, W$ will serve for Theorem 4.8. Now we rewrite the ω-character π as a p-character $< \zeta, k_{p-1}, \ldots, k_0 >$, and we evaluate $F_0^{k_0} \cdots F_{p-1}^{k_{p-1}} F_p^{\zeta}[a]$. This value belongs to W just in case Σ holds in $[\alpha,<]$, α countable and of ω-character π. Q.E.D.

<u>Theorem</u> 4.12: For any countable ordinal α, $MT[\alpha,<]$ is decidable. The decision method depends on the ω-character of α, only. The theory MT(co) of all countable ordinals is decidable.

Proof: The first part clearly follows from 4.12. The following indicates a decision method for MT(co). Given the MT-sentence Σ, construct the absorption automaton $\mathbf{A} = [K, a, F_0, \ldots, F_p, W]$ as in the proof of 4.12. Now Σ belongs to MT(co) just in case $\mathbf{A}$ accepts all p-characters $< \zeta, k_{p-1}, \ldots, k_0 >$. It is easy to make up a procedure for deciding whether or not $\mathbf{A}$ accepts all inputs (see automata theory). Q.E.D.

We have now listed the main conclusion coming out of Theorem 4.5, for sentences. The trick was, to reformulate input-free finite-state recursions as absorption automata, which work on characters rather than ordinals, and which are very simple gadgets. It remains to use Theorem 4.5 to study definability in the MT of countable ordinals, i.e., to investigate the ω_1-behavior of finite automata with more than one input-state. We have not investigated this matter; one should try to extend the notion of absorption-automata from the input-free to the general case. Definability in $MT[\omega_0, <]$ is extensively discussed in [15]. Also see [6].

5. The filters of cofinal closed sets.

We begin with an analysis of what goes wrong at $\varkappa_1$ (i.e., ω_1 if AC_{ω}^{ω}), with our proof in Section 4. First of all, we have nicely isolated the trouble to (4) (which contains the rudiments of a complementation-lemma, and is apart from the more intricate details of the subset-construction). In the second part of the proof of (4), we used that x is an ω_0-limit, so the splicing places y_i could be made to form a simple ω_0-sequence approaching x. At $x = \varkappa_1$, for the first time, such a ω_0-approach is not available. We now must take care that the result of splicing the pieces $z[y_i, y_{i+1}]$ becomes a $\varkappa_1$-sequence. That is, we must choose the y_i to be closed under limit. We will now show that this is not possible, without drastic revisions in the formula (4). In fact, we will show that the complementation lemma, for the transition-systems used in Section 3, is false at $\varkappa_1$. Since the other parts of the proof work for $\varkappa_1$, also (4) must fail at $x=\varkappa_1$.

For the purpose of finding our counterexample, and for the purpose of revising (4), it is natural to investigate the notion of cofinal-closed subset of $\varkappa_1$. By closed we actually mean closed relative to $\varkappa_1$, that is, the limit $\varkappa_1$ is not counted in the closure of $X \subseteq \varkappa_1$ under limits. We need the following important fact.

Lemma 5.1: If X_i, $i<\omega_0$ are cofinal-closed subsets of $\varkappa_1$, then so is $X = \bigcap_{i<\omega_0} X_i$.

Proof: As an intersection of closed sets is closed, we only have to show that X is cofinal. Let $u<\varkappa_1$. We are to show an element x of X, such that $u<x$. For this purpose we define $x_{i,j}$ by induction over $i,j<\omega_0$:

$$x_{0,0} = u \qquad u=x_{0,0}<\varkappa_1$$
$$x_{i+1,j} = (\mu t).t\in X_i \wedge x_{i,j}<t \text{ exists as } X_i \text{ cofinal; } \quad x_{i,j}<x_{i+1,j}<\varkappa_1, x_{i+1,j}\in X_i$$
$$x_{0,j+1} = \lim_{i<\omega_0} x_{i,j} \qquad \text{as } \varkappa_1 \text{ is not } \omega_0\text{-limit, } x_{i,j}<x_{0,j+1}<\varkappa_1$$

Now define $x=\lim_{j<\omega_0} x_{0,j}$. We clearly have $u<x$. As $x_{0,j}<x_{i+1,j}<x_{0,j+1}$, we have $x = \lim_{j<\omega_0} x_{i+1,j}$. Therefore, as $x_{i+1,j}\in X_i$ and X_i is closed, $x\in X_i$. As this goes for all $i<\omega_0$, we thus have found our $x\in X$, $u<x$. Q.E.D.

We now define the class $\mathcal{J}_1^\alpha$ of all those subsets of α which contain a cofinal closed set, and $\mathcal{O}_1^\alpha = \{\tilde{X}; X\in\mathcal{J}_1^\alpha\}$, the class of all those subsets of α, which do not meet some cofinal-closed set. If $\alpha=\varkappa_1$ we will often use the shorter notations $\mathcal{J}_1$ and $\mathcal{O}_1$. Thus,

(1) $\quad X\in\mathcal{J}_1 \ : \ X\subseteq\varkappa_1 \wedge (\exists Q)[Q\subseteq X \wedge Q \text{ cofinal-closed in } \varkappa_1]$

Theorem 5.2: The class $\mathcal{J}_1$ is a ω_0-filter on $\mathcal{P}\varkappa_1$. That is, $\mathcal{J}_1$ is closed under ω_0-intersection, and any superset of $X\in\mathcal{J}_1$ still belongs to $\mathcal{J}_1$. Equivalently, $\mathcal{O}_1$ is a ω_0-ideal in the Boolean algebra of all subsets of $\varkappa_1$.

This is clearly a corollary to Lemma 5.1. It is easy to see that both, the lemma and the theorem, generalize to any $\varkappa_1$-limit α. In fact, a stronger form holds for any limit α, which is not an ω_0-limit. There is a second remark which will be important for our work. If X is a set of ordinals, let X' (<u>the derivative of</u> X) denote the set of all limits of members of X. So $X \cup X'$ is the closure of X, under limit. As usual, X' is closed.

<u>Caution</u>!

In a context, in which we deal only with subsets of α, we use the notation X' in the relativized sense, just as we have used "closed" in the sense relative α. That is, α is not to be counted to X', even if X is cofinal α.

<u>Remark</u> 5.3: A subset X of $\varkappa_1$ is cofinal if and only if its derivative X' is cofinal, and hence is cofinal-closed. Also, X is cofinal $\varkappa_1$ just in case $X' \in \mathcal{J}_1$.

<u>Proof</u>: The only thing not obvious here is: X' is cofinal if X is cofinal (in fact AC would have to be used, if we replace $\varkappa_1$ by ω_1, and X' can be empty if $\varkappa_1$ is replaced by an ω_0-limit α). Suppose then that X is cofinal $\varkappa_1$, and let $u<\varkappa_1$. We are to find $x \in X'$, such that $u<x$. For this purpose define $x_0=u$ and $x_{i+1} = (\mu t)[t \in X \wedge x_i<t]$. As X is cofinal, all x_i for $i<\omega_0$ exist. Clearly $x_i \in X$. Let $x=\lim_{i<\omega_0} x_i$. As $\varkappa_1$ is not ω_0-limit, we have $u<x<\varkappa_1$. So x is a limit of

elements of X, and $x \neq \aleph_1$. Therefore $u < x \in X'$. Q.E.D.

If $\{t; Zt=a\} \in \mathcal{J}_1$, i.e., if there is a cofinal closed Q such that $Qt \supset Zt=a$, then we say that a <u>occurs</u> <u>cofinal-closed</u> as state of Z, or that a <u>occurs</u> $\mathcal{J}_1$ <u>times</u> as state of Z. The following may be construed as a <u>weak</u> <u>Ramsey-lemma</u> for $\aleph_1$.

<u>Remark</u> 5.4: Let Z be any $\aleph_1$-sequence from a finite set of states. At most one state of Z occurs cofinal-closed in $\aleph_1$. Exactly one set D of states occurs cofinal closed as $\sup^x Z$, namely $D = \sup^{\aleph_1} Z$.

<u>Proof</u>: The first claim is just a restatement of "$\mathcal{J}_1$ is a filter". Now let $D = \sup^{\aleph_1} Z$. To show the second claim it remains only to exhibit a cofinal-closed Q of limits, such that $\sup^x Z = D$ for any $x \in Q$. By the definition of D we clearly have a cofinal set P, such that
$v<y \wedge Pv \wedge Py . \supset . \{Zt; v \leq t < y\} = D$. Now, for every $x \in P'$ the set D keeps occurring cofinal x, as values of Z. So, $\sup^x Z = D$, for any $x \in P'$. As P' is cofinal-closed (Remark 5.3), it will serve as Q. Q.E.D.

We are now ready to provide the counterexample announced at the beginning of this section. First note that the notion "X is cofinal-closed" is MT-definable by the formula $(\forall x)[(\exists^x t)Xt \supset Xx] \wedge (\exists^{\aleph_1} t)Xt$. Therefore we have,

(2) $X \in \mathcal{J}_1 \quad .\equiv. \quad (\exists Z).(\forall t)[Zt \supset Xt] \wedge (\forall x)[(\exists^x t)Zt \supset Zx] \wedge (\exists^{\aleph_1} t)Zt$

Note that the transition-condition $(\forall x)[---]$ for limits is of the form $\mathcal{A}[\sup^x Z, Zx]$, and the terminal condition is of form $\mathcal{L}[\sup^{\aleph_1} Z]$. Hence (2) shows that $\mathcal{J}_1$ is expressible (as accepted set) by a sup-transition-system, of the type we used in Section 4. We now show that the same can hold for $\sim \mathcal{J}_1$, only in a very strange set-theory.

<u>Remark</u> 5.5: The complement $\sim \mathcal{J}_1$ is expressible by a sup-transition-system, if and only if, the filter $\mathcal{J}_1$ is prime.

<u>Proof</u>: If $\mathcal{J}_1$ is prime we have $X \notin \mathcal{J}_1 \equiv \tilde{X} \in \mathcal{J}_1$. Thus, replacing X by $\tilde{X}$ in the right side of (2) yields a definition for $\sim \mathcal{J}_1$. So $\sim \mathcal{J}_1$ is defined by a transition-system. Suppose now that there is a transition-system for $\sim \mathcal{J}_1$. So we have E, H, $\mathcal{K}$, $\mathcal{L}$ such that,

(a) $X \notin \mathcal{J}_1 \;.\equiv.\; (\exists Y).E[Y0] \wedge (\forall t)H[Xt, Yt, Yt'] \wedge (\forall x)\mathcal{K}[\sup^x Y, Yx] \wedge \mathcal{L}[\sup^{\aleph_1} Y]$

Assume $\tilde{X} \notin \mathcal{J}_1$. We are to show that $X \in \mathcal{J}_1$. Because of (a) this is accomplished if we show: For any X-run Y of E, H, $\mathcal{K}$, and $D = \sup^{\aleph_1} Y$ we have $\sim \mathcal{L}[D]$. The proof of this goes as follows.

Our assumptions are: $\tilde{X} \notin \mathcal{J}_1$, Y is an X-run of E, H, $\mathcal{K}$, $D = \sup Y$ at $\aleph_1$. By Remark 5.4 there is a cofinal-closed Q, such that $\sup^z Y = D$, for all $z \in Q$. We may assume that $\{Yt; z<t\} = D$ for all $z \in Q$. As $\tilde{X} \notin \mathcal{J}_1$ and Q is cofinal-closed, there is an x such that $x \in Q$ and $x \in X$. Let $d = Yx$. As Y is a run of $\mathcal{K}$ and $\sup^x Y = D$, we have $\mathcal{K}[D, d]$. As $\{Yt; x<t\} = D$

and $d \in D$, there is a $y > x$, such that $Yy = d$ and $\{Yt; x \leq t < y\} = D$. Now we make up X_1 by starting with the initial segment $X[0,x)$, and then repeating the same segment $X[x,y)$ until we have a $\varkappa_1$-string (i.e., $\varkappa_1$ times). In exactly the same manner we make up Y_1 from the phase $Y[0,x)$ and the period $Y[x,y)$. Note that the splicing points form the set $P = \{x+ut; t<\varkappa_1\}$, whereby $y=x+u$. Clearly X_1, Y_1 satisfy the initial condition E, and the transition-condition H, $\mathcal{K}$, for places not in P, and places $x+u(t+1)$. But at limits $x+u \cdot z$, we have $\sup Y_1 = D$ and $Y_1 = d$. As $\mathcal{K}[D,d]$ we thus see that Y_1 is an X_1-run of E, H, $\mathcal{K}$ and clearly $\sup^{\varkappa 1} Y_1 = D$. Because x belongs to X, the truth value T occurs in the sequence X_1, at every place $x+ut \in P$. Hence, $P \subseteq X_1$ and as P clearly is cofinal-closed in $\varkappa_1$, we have $X_1 \in \mathcal{J}_1$. By (a) and the fact that Y_1 is an X_1-run with $\sup Y_1 = D$, we conclude that $\sim \mathcal{L}[D]$. Q.E.D.

Suppose now AC. In [17] Ulam has shown that no non-principal ω_0-filter in $\omega_1 = \varkappa_1$ can be prime. Hence by 5.2 and 5.3, $\sim \mathcal{J}_1$ is not expressible by a sup-transition-system. But $\mathcal{J}_1$ is, see (2). Hence, the complementation-lemma at ω_1 will not work for sup-systems. In fact, this will not work under the weaker assumption: $\mathcal{J}_1$ is not prime.

Let us now consider for a moment the strange case: $\mathcal{J}_1$ is prime. It is not likely that the proof of (4) in Section 3 can be fixed up. However, if (4) is modified as in Section 6, it is easy to modify the proof in Section 6, using "$\mathcal{J}_1$ is prime", to obtain a complementation-lemma for $\sup_1$-systems. But, as now

$\mathcal{J}_1$ and $\sim\mathcal{J}_1$ are expressible by sup-systems, it follows that the complementation-lemma holds for sup-systems.

The proof, which we present in Section 6, actually uses the assumption that the quotient $\mathcal{P}\kappa_1/\mathcal{J}_1$ has no atoms. We will now show that Ulam's argument does really yield this stronger consequence of AC_ω^γ $(\gamma=2^\omega)$. As in particular AC_ω^ω is assumed we have $\kappa_1=\omega_1$. It should be noted that Ulam's proof very much uses the cardinal definition of ω_1, rather than the accessibility definition.

Theorem 5.6 (AC): Let $\mathcal{J}$ be an ω_0-filter on $\mathcal{P}\omega_1$, which contains all terminal segments of ω_1. Every non-null element of the quotient $\mathcal{P}\omega_1/\mathcal{J}$ has breadth ω_1. That is, in any $U\notin\mathcal{O}$ there are ω_1 many disjoint $U_i\notin\mathcal{O}$. $\mathcal{O}$ denotes the ideal corresponding to $\mathcal{J}$.

Proof: Using AC_ω^γ we know there is a function f defined on ω_1, such that, for any $x<\omega_1$, f_x is a one-one map of x into ω_0. With Ulam we define the set S_x^i, for $i<\omega_0$ and $x<\omega_1$:

$$y\in S_x^i \;.\equiv.\; x<y \wedge f_y x=i$$

It is easy to show (a) $x<\omega_1 \supset \bigcup_{i<\omega_0} S_x^i = (x,\omega_1)$, and (b) $i<\omega_0 \wedge$ $\wedge\, u<x<\omega_1 .\supset. S_u^i \cap S_x^i = 0$.

Assume $U\subseteq\omega_1$, $U\notin\mathcal{O}$. By our assumptions $(x,\omega_1)\in\mathcal{J}$. Therefore, $U\cap(x,\omega_1)$ is equivalent to U, modulo $\mathcal{J}$. So $U\cap(x,\omega_1)\notin\mathcal{O}$. By (a) and because $\mathcal{O}$ is an ω_0-ideal, this implies $(\forall x)^{\omega_1}(\exists i)^{\omega_0} U\cap S_x^i\notin\mathcal{O}$. It follows that there exists an $i<\omega_0$, and cofinal ω_1 many x, such that $U\cap S_x^i\notin\mathcal{O}$. Thus we have found

ω_1 many subsets $U_x = U \cap S_x^i \notin \mathcal{O}$ of U, which by (b) are mutually disjoint. Q.E.D.

<u>Corollary</u> 5.7 (AC): The Boolean algebra $\mathcal{P}\omega_1/\mathcal{J}_1$ has no atom.

Using AC_ω^ω we know $\omega_1=\aleph_1$, so by Theorem 5.2, $\mathcal{J}_1$ is an ω-filter. It clearly contains all terminal segments. Hence, the corollary follows from Theorem 5.6.

In Section 6 we will only need the corollary. So, Theorem 5.6 is both much too strong and much too general for our purpose. The perplexing thing is that, even though our filter $\mathcal{J}_1$ is a very natural one, we do not know a more constructive proof of Corollary 5.7.

<u>Problem</u>: Without AC, can one show that $\mathcal{J}_1$ on $\aleph_1$ is (a) not prime, (b) admits no atom? Is it possible that $\mathcal{J}_1$ is not prime, but admits an atom?

In future work on ω_2 we intend to use the filter $\mathcal{J}_2$ of those sets, which contain a cofinal set which is closed under ω_1-limits. We will use correspondingly finer filters $\mathcal{J}_\alpha$ at limits of higher accessibility (not cardinality!). Note that $\mathcal{J}_{\alpha+1}$ is an ω_α-filter, and that Ulam's argument trivially extends to ω_α, $\omega_{\alpha+1}$ in place of ω_0, ω_1.

Let $\mathcal{J}$ be any filter on the set A, let $\mathcal{O} = \{\tilde{U}; U \in \mathcal{J}\}$ be the corresponding ideal. The following notations are very convenient, as they can be handled in much the same way as the standard quantifiers with range A. In fact, these correspond to the trivial filter $\mathcal{J} = \{A\}$.

$$(3) \qquad \begin{array}{lcl} (\forall_{\mathcal{J}}x)Ux & : & U\in\mathcal{J} \\ (\exists_{\mathcal{J}}x)Ux & : & U\notin\mathcal{O} \\ \sup_{\mathcal{J}}Z & : & \{c;(\exists_{\mathcal{J}}x)Zx=c\} \end{array}$$

Here Z denotes an n-tuple of unary predicates, or more generally, a map from A into some finite set K. The reader will verify the usual laws for distributing quantifiers over conjunctions and disjunctions. He will also establish the following alternative definition of $\sup_{\mathcal{J}}$.

$$(4) \qquad \sup_{\mathcal{J}}Z = \text{the smallest } D, \text{ such that } (\forall_{\mathcal{J}}x)Zx\in D$$

Note that, in the standard case $\mathcal{J} = \{A\}$, supZ just becomes the set valZ, of values taken on by Z.

In case $A = \alpha$ is a limit ordinal (or a linear order without last element), we have the filter $\mathcal{J}_0^\alpha$ consisting of all $U\subseteq\alpha$ which are terminal in α(i.e., contain a terminal segment of α). Note that our notations $(\exists^\alpha t)$, $(\forall^\alpha t)$, $\sup^\alpha$, used throughout Section 4, are just the quantifiers (3) in the case $\mathcal{J} = \mathcal{J}_0^\alpha$. In Section 6, we will use the more explicit notations $(\exists_0^\alpha t)$, $(\forall_0^\alpha t)$, and $\sup_0^\alpha$, when referring to the terminal filter $\mathcal{J}_0^\alpha$. It is interesting to note that this filter, too, is a filter of cofinal-closed sets. Only now the successor operation is used in addition to lim, in the closure-process.

In case $A = \alpha$ is a limit, but not an ω_0-limit, we have just introduced the refinement $\mathcal{J}_1^\alpha$, of the filter $\mathcal{J}_0^\alpha$. When referring to this filter $\mathcal{J}_1^\alpha$, we will use the abbreviated notations $\exists_1^\alpha$, $\forall_1^\alpha$, $\sup_1^\alpha$, in place of the notations (3). We now present the general proof of Lemma 1.2.

<u>Lemma</u> 5.8: Let $\mathcal{J}$ be any filter on A. a) Every condition of either form $(\exists_{\mathcal{J}}t)L[Zt]$, $(\forall_{\mathcal{J}}t)L[Zt]$ can be stated in the form $\mathcal{K}[\sup_{\mathcal{J}}Z]$. b) A disjunction of conditions $\mathcal{K}_1[\sup_{\mathcal{J}}Z_1]$, $\mathcal{K}_2[\sup_{\mathcal{J}}Z_2]$ can be stated in the form $\mathcal{K}[\sup_{\mathcal{J}}Z_1Z_2]$. c) Every condition $\mathcal{K}[\sup_{\mathcal{J}}Z]$ can be stated as a Boolean expression in parts of form $(\exists_{\mathcal{J}}t)L[Zt]$. Hence, $\sup_{\mathcal{J}}$-conditions are exactly the Boolean expressions in $\exists_{\mathcal{J}}$-conditions (i.e., $\forall_{\mathcal{J}}$-conditions).

<u>Proof</u>: We drop the index $\mathcal{J}$. From our definition (3) we clearly have $(\exists t)L[Zt]$ just in case $L \cap \sup Z \neq 0$, and by (4), $(\forall t)L[Zt]$ just in case $\sup Z \subseteq L$. This proves a).

Now let $\mathcal{K}_1$ be a set of subsets of the possible states of Z_1 ($\{F,T\}^n$ if Z_1 is an n-tuple), and let $\mathcal{K}_2$ be such a set for Z_2 (which may be an m-tuple, $m \neq n$). We define the set $\mathcal{K}$ to consist of all those sets D of pairs ab (a a state of Z_1, b a state of Z_2), such that the two projections of D satisfy $D_1 \in \mathcal{K}_1 \vee D_2 \in \mathcal{K}_2$. To show b) holds for this $\mathcal{K}$, it evidently suffices to see that $\sup Z_1$ is the first projection of $\sup Z_1Z_2$, and $\sup Z_2$ is the second projection. The reader will verify this, using (3, 4).

Clearly we have $\mathcal{K}[\sup Z]$ just in case

$$\bigvee_{D\in\mathcal{K}} .(\forall t)D[Zt] \wedge \bigwedge_{e\in D} (\exists t)Zt=e.$$

This takes care of c). Q.E.D.

6. A decision method for $MT[\omega_1,<]$

We will here take as a model our new proof of the complementation lemma at ω_0. In that proof the filter $\mathcal{J}_0$ of terminal subsets of ω_0 plays a fundamental role. We have seen that these filters $\mathcal{J}_0^{\alpha}$ work well at all ω_0-limits α, and in Section 5 we have seen what goes wrong at $\alpha=\varkappa_1$. From this analysis it is obvious that the finer filter $\mathcal{J}_1$ will have to be used in transitions to $\varkappa_1$-limits. We will show here that $\sup_1$-terminal-conditions will do for a complementation lemma at $\varkappa_1$. We will be using AC_{ω}^{γ}, so $\omega_1=\varkappa_1$, and $\mathcal{P}\omega_1/\mathcal{J}_1$ has no atom.

All our results, up to the statement of the complementation lemma, are about the following transition-system with terminal condition:

$$(1) \quad \Gamma(X,Z) \quad : \quad E[Z0] \wedge (\forall t)H[Xt,Zt,Zt'] \wedge (\forall x)\mathcal{K}[\sup_0^x Z,Zx] \wedge \mathcal{L}[\sup_1 Z]$$

Here or in the sequel, the superscript ω_1 (or $\varkappa_1$) is meant to be attached to such expressions as $\sup_1$, $\exists_1$, $\forall_1$. By $\Gamma_u^v(X,Z)$ we mean the $H,\mathcal{K}$-system, resulting from Γ by dropping the initial- and terminal-condition, and by restricting $(\forall t)$ to $(\forall t)_u^v$, and $(\forall x)$ to $(\forall x)_u^{v'}$. So Γ_u^v is just the system we investigated in Section 4. From it the subsets $S_u^{c,C}v$ are defined, just as in Section 4:

(2) $S_u^{c,C}v$: $\{b; \exists Z[u,v].Zu=c \wedge \Gamma_u^v(X,Z) \wedge Zv=b \wedge \{Zt; u \leq t < v\} = C\}$

The abbreviation $S_u v$ is used for the vector of all $S_u^{c,C}v$, and states of sequences of type S_u are denoted by s, s_0, etc. The fact, shown in Section 4, that for any ω_1-sequence X, the sequences S_u can be obtained by a finite-state $\sup_0$-recursion (subset-construction), will be used extensively. A first use is in the following consequence of our weak Ramsey-lemma in Section 5.

Lemma 6.1: Let X be any input (of length ω_1), let $u<\omega_1$. There is a unique state s_u, such that $(\forall_1 y) S_u y = s_u$. Furthermore, $s_u = s_v$ if and only if $u \approx v(-\omega_1)$.

Proof: The subset-construction, Lemma 4.4, is an ω_1-recursion of form $Y0=e$, $Yt' = F[Xt,Yt]$, $Yx = \mathcal{S}[\sup_0^x Y]$, and S_0 is one of the components of Y. By Remark 5.4, some set D of states of Y occurs cofinal closed as $\sup_0^x Y$, i.e., $(\forall_1 x)\sup_0^x Y = D$. Hence, if $d = \mathcal{S}[D]$, we have $(\forall_1 x) Yx = d$. As S_0 is one of the components of Y, and if s_0 is the corresponding component of d, we therefore have $(\forall_1 x) S_0 x = s_0$. As $\mathcal{J}_1$ is a filter, this s_0 is unique. Now, S_u is defined from $X[u,\omega_1)$ in exactly the same way as S_0 is defined from $X = X[0,\omega_1)$. Hence, the first part of our lemma is proved. The second part is a simple consequence of the fact that $\mathcal{J}_1$ is a filter, and contains all terminal segments of ω_1. Q.E.D.

We intend to find a transition-system Γ', which is of the same form as Γ, and accepts the complementary set of inputs.

The transition part of Γ' will consist, in essence, of the subset-recursion in Section 4 (as in Section 3 this part of Γ' was made up from the subset-recursion for finite ordinals). We now introduce the formulas Ω, which will become the terminal-condition of Γ'.

(3) $$\Omega_{s_0,T} : \bigvee_{\substack{W \\ w \to s_w}} . \bigwedge_{w \in W} (\exists_1 y) y \approx w(-\omega_1) \wedge (\forall_1 y) \bigvee_{w \in W} y \approx w(-\omega_1) \wedge (\forall_1 y) S_0 y = s_0 \wedge \bigwedge_{w \in W} (\forall_1 y) S_w y = s_w$$

Here T is any set of states of the type s. The notation $w \to s_w$ means a one-to-one map from W onto T. So, the first disjunction symbol is to be read: There is a finite set W of ordinals $< \omega_1$, and there is a one-one map $w \to s_w$ from W onto T. The motivation for (3) comes from the corresponding formula (6) of Section 3. Note that the first two conditions on W make it a system of representatives of exactly those classes modulo $\approx (-\omega_1)$, which occur cofinal-closed in ω_1. That is, up to choice of representatives, W is required to be $\sup_1$ of the classes of $\approx (-\omega_1)$. This uniqueness of W makes the situation at ω_1 actually nicer than that at ω_0. Also note here, in contrast to ω_0, the quantifications refer separately to $y \approx w(-\omega_1)$ and $S_w y = s_w$. The following lemma is the analog to 3.3.

<u>Lemma</u> 6.2: For every X there are s_0 and T, such that $\Omega_{s_0,T}$.

<u>Proof</u>: We know that $\approx(-\omega_1)$ is an equivalence of finite index. Some of the classes will come up cofinal ω_1, and some of these will occur non-null in the sense of $\mathcal{F}_1$. Let W be a system of representatives for the latter classes. So we clearly have, $w\in W \supset (\exists_1 y) y \approx w(-\omega_1)$ and $(\forall_1 y) \bigvee_{w\in W} y \approx w(-\omega_1)$. By our Ramsey-result, Lemma 6.1, we have s_0 and s_w, for each $w\in W$, such that $(\forall_1 y) S_0 y = s_0$ and $(\forall_1 y) S_w y = s_w$. We now set $T=\{s_w; w\in W\}$. Now, s_0 and T will clearly be as required in our lemma, if we can show that $w\to s_w$ maps W one-one onto T. By definition of T it is onto. Suppose now $w,v\in W$ and $s_w=s_v$. Then, by Lemma 6.1, $w \approx v(-\omega_1)$. As W is a partial system of representatives modulo $\approx(-\omega_1)$, this implies $w=v$. So we have shown that $w\to s_w$ is also one-one. Q.E.D.

We now restate our Ω in a slightly modified form.

(4)
$$\Omega_{s_0,T} .\equiv. (\exists P_s)_{s\in T}. (\forall y) \bigwedge_{s\in T} [P_s y \supset S_0 y = s_0] \wedge (\forall y)(\forall v)^{y} \bigwedge_{s,r\in T} [P_s v \wedge P_r y \supset S_v y = s]$$
$$\wedge \bigwedge_{s\in T} (\exists_1 y) P_s y \wedge (\forall_1 y) \bigvee_{s\in T} P_s y$$

<u>Proof</u>: Suppose first that $\Omega_{s_0,T}$. So, by definition (3) there are a W, and a one-one map $w\to s_w$ from W onto T, and sets P_w such that (a) $w\in W \supset P_w \not\in \mathcal{O}_1$, (b) $w\in W \wedge P_w y .\supset. y \approx w(-\omega_1)$, (c) $\bigcup_{w\in W} P_w = P \in \mathcal{F}_1$, $(\forall_1 y) S_0 y = s_0$, and $w\in W \supset (\forall_1 y) S_w y = s_w$. As $\mathcal{F}_1$ is a filter and W is finite, the last three statements imply that the part $\overline{P}$ of P, for which $S_0 y = s_0$ and $S_w y = s_w$, will still be in $\mathcal{F}_1$. Also, the new forms $\overline{P} \cap P_w$ will still have all the properties of the old P_w. We may therefore assume now that

(d) $Py \supset S_0 y = s_0$, and (e) $w \in W \wedge Py \supset S_w y = s_w$. Suppose now that $v, w \in W$ and $y \in P_v \cap P_w$. Then, by (b), $v \approx w(-\omega_1)$. Because P is cofinal ω_1, this implies existence of $t \in P$ such that $S_v t = S_w t$. Hence by (e), $s_v = s_w$. As $w \to s_w$ is one-one, we therefore have shown that $v, w \in W$ and $y \in P_v \cap P_w$ implies $v = w$. I.e., (f) $v \neq w \in W . \supset . P_v \cap P_w = 0$. By (c) there is a cofinal-closed $Q \subseteq P$. Clearly Q and the sets $Q \cap P_w$ will have all the stated properties of P and the P_w. We may therefore assume (g) P is cofinal-closed. We now define y_i, w_i, t_i by the following induction over $i < \omega_1$:

$y_0 = (\mu y) Py$	exists by (g)	;	Py_0
$w_i = (\mu w) . w \in W \wedge P_w y_i$	exists as Py_i and (c)	;	$P_{w_i} y_i$, $w_i \in W$, by (b), $y_i \approx w_i(-\omega_1)$
$t_i = (\mu t) . y_i \approx w_i(t)$	exists as $y_i \approx w_i(-\omega_1)$	;	$y_i \approx w_i(t_i)$
$y_{i+1} = (\mu y) . Py \wedge t_i < y$	exists by (g)	;	Py_{i+1}, $t_i < y_{i+1}$, hence $y_i \approx w_i(y_{i+1})$
$y_\lambda = \lim_{i<\lambda} y_i$	ω_1 is not ω_0-limit	;	Py_λ, as Py_i and (g)

Now clearly $\overline{P} = \{y_i\}$ is cofinal closed, and hence in $\mathcal{J}_1$. Let $\overline{P}_w = \overline{P} \cap P_w$. By (a, c, d) we clearly have, (h) $w \in W \supset \overline{P}_w \notin \mathcal{O}_1$, $\bigcup_{w \in W} \overline{P}_w \in \mathcal{J}_1$, and $w \in W \wedge \overline{P}_w y \supset S_0 y = s_0$.

Suppose now that $u, w \in W$, $v < y$, $\overline{P}_u v$, and $\overline{P}_w y$. Then $\overline{P}v$ and $\overline{P}y$, and by the definition of $\overline{P}$, there are $i < j$ such that $v = y_i$ and $y = y_j$. As $y_i \approx w_i(y_{i+1})$ and $y_{i+1} \leq y_j$, we therefore have $v \approx w_i(y)$. As $v = y_i \in P_{w_i}$ and $v \in \overline{P}_u$, we have by (f), $w_i = u$. Hence, $v \approx u(y)$, i.e., $S_v y = S_u y$. As $y \in \overline{P}_w \subseteq P$ we have by (e), $S_u y = s_u$. Hence, $S_v y = s_u$. This argument shows

(k) $v<y \wedge u, w \in W \wedge \overline{P}_u v \wedge \overline{P}_w y .\supset. S_v y = s_u$. Now we recall that the map $w \to s_w$ is one-one from W onto T. The sets P_s, $s \in T$ are therefore well defined by the formula $P_{s_w} = \overline{P}_w$. Now (h) and (k) show the properties required for the P_s, on the right side of (4). This ends the proof of (4) from left to right.

Assume now the right side of (4). So there are sets P_s such that (m) $s \in T \supset P_s \notin \mathcal{O}_1$, (n) $P = \bigcup_{s \in T} P_s \in \mathcal{J}_1$, (p) $Py \supset S_0 y = s_0$, and (q) $s \in T \wedge v<y \wedge P_s v \wedge Py .\supset. S_v y = s$. From (n) and (p) we have (r) $(\forall_1 y) S_0 y = s_0$. Now define w_s to be the first member of P_s, and $W = \{w_s; s \in T\}$. From (q) we clearly have, (s) $s \in T \wedge w_s < y \in P .\supset. S_{w_s} y = s$.

Suppose $s, r \in T$ and $w_s = w_r = w$. By (n) there is a $y \in P$, $w<y$. By (s) it follows $s = S_w y = r$. This shows that $s \to w_s$ maps T one-one onto W. Let $w \to s_w$ be the converse to this map. So we have (t) $w \to s_w$ maps W one-one onto T. Furthermore, if we set $P_w = P_{s_w}$, we can restate (m, n, q, s) as $(\overline{m})$ $w \in W \supset P_w \notin \mathcal{O}_1$, $(\overline{n})$ $P = \bigcup_{w \in W} P_w \in \mathcal{J}_1$, $(\overline{q})$ $w \in W \wedge v<y \wedge P_w v \wedge Py \supset S_v y = s_w$, and $(\overline{s})$ $w \in W \wedge w<y \in P \supset S_w y = s_w$. From $(\overline{n}, \overline{s})$ we see (u) $w \in W \supset (\forall_1 y) S_w y = s_w$.

Suppose $w \in W$ and $P_w v$. Because of $(\overline{m})$ there is a $y \in P_w$, $v<y$. Therefore by $(\overline{q})$, $S_v y = s_w$. As w is the first member of P_w, we also have $w<y$. Also $y \in P$. Hence by $(\overline{s})$, $S_w y = s_w$. Therefore, $S_v y = S_w y$, which implies $v \approx w(-\omega_1)$. This argument shows $w \in W \wedge P_w v \supset v \approx w(-\omega_1)$. Hence by $(\overline{n})$ we have (v) $(\forall_1 y) \bigvee_{w \in W} y \approx w(-\omega_1)$, and by $(\overline{m})$ we have (w) $w \in W \supset (\exists_1 y) y \approx w(-\omega_1)$. Now (t, w, v, r, u) are the requirements on W and the map $w \to s_w$, occurring on the

right side of the definition (3). Hence we have proved $\Omega_{s_0,T}$. Q.E.D.

We now come to the heart of our proof of the complementation lemma. Define,

$$B[s_0,T] \;:\; \bigvee_{c,C,D_0,D_1} .E[c] \wedge \bigwedge_{e\in D_1} \mathcal{K}[D_0,e] \wedge \mathcal{L}[D_1] \wedge D_1 \subseteq s_0{}^{c,C} \wedge \tag{5}$$

$$\bigvee_{Q\subseteq D_1\times T} [Q \text{ onto } T \wedge Q \text{ onto } D_1 \wedge \bigwedge_{e,s\in Q} D_1 \subseteq s^{e,D_0}]$$

<u>Lemma</u> 6.3: For any input X, and any s_0, T, if $\Omega_{s_0,T}$ then $(\exists Z)\Gamma(X, Z) .\equiv. B[s_0,T]$.

<u>Proof</u>: Assume $\Omega_{s_0,T}$. By (4) we have sets P_s such that (a) $s\in T \wedge P_s y \supset S_0 y = s_0$, (b) $s,r\in T \wedge v<y \wedge P_s v \wedge P_r y \supset S_v y = s$, (c) $s\in T \supset P_s \notin \mathcal{O}_1$, (d) $P_T = \bigcup_{s\in T} P_s \in \mathcal{J}_1$. Furthermore, we may assume that P_T is cofinal-closed (by (d) there is a subset Q of P_T which has this property, and Q together with the $Q \cap P_s$ has all the stated properties of P_T and P_s).

Suppose now $(\exists Z)\Gamma(X,Z)$. So there is a Z, such that $E[Z0]$, Z is an X-run of H and $\mathcal{K}$, and $\mathcal{L}[\sup_1 Z]$. Let $c=Z0$, $D_0 = \sup_0 Z$, and $D_1=\sup_1 Z$. So we have (e) $E[c]$ and $\mathcal{L}[D_1]$. By the definition of D_1, there are sets P_e, such that (f) $e\in D_1 \supset P_e \notin \mathcal{O}_1$, (g) $P = \bigcup_{e\in D_1} P_e \in \mathcal{J}_1$, and (h) $e\in D_1 \wedge P_e y \supset Zy=e$. We may assume in addition that P is cofinal-closed (else select subset which has this additional property). Now we define y_i by induction on $i<\omega_1$:

$y_0 = (\mu y).Py \wedge D_0 \subseteq \{Zt; 0 \leq t < y\} \wedge (\forall t)_y Zt \in D_0$	exists as P is cofinal and $D_0 = \sup_0 Z$
$y_{i+1} = (\mu y).Py \wedge y_i < y \wedge D_0 = \{Zt; y_i \leq t < y\}$	exists for same reason
$y_\lambda = \lim_{i<\lambda} y_i$	exists as ω_1 is not ω_0-limit

Note that $y_i \in P$; in particular this holds at limits i, because P is closed. As $y_i < y_{i+1}$ we thus have a cofinal-closed subset $\{y_i\} \subseteq P$. For the old reason, we may therefore assume that $P = \{y_i\}$, without destroying the stated properties of P; the new P_e's being $\{y_i\} \cap P_e$. So we now have the additional facts (i) $v<y \wedge Pv \wedge Py \supset \{Zt; v \leq t < y\} = D_0$, and (j) $Py \supset \{Zt; 0 \leq t < y\} = C$, whereby $D_0 \subseteq C = \{Zt; 0 \leq t < y_0\}$. Because of (i) we clearly have $\sup_0^x Z = D_0$, for every $x \in P'$ (the derivative of P). Note that P' still is cofinal-closed, and $P' \subseteq P$, so we can choose P' as new P and redefine the P_e. Without losing the old facts we now have $Px \supset \sup_0^x Z = D_0$. As Z is a run of $\mathcal{K}$, and by (g, h), we therefore see, (k) $e \in D_1 \supset \mathcal{K}[D_0, e]$.

Suppose $e \in D_1$. By (d, f) we have $P_T \cap P_e \notin \mathcal{O}_1$, and therefore there is a $y \in P_T \cap P_e$. Because of (j, g), $\{Zt; 0 \leq t < y\} = C$, and by (h), $Zy = e$. Hence, as Z is a run of H, $\mathcal{K}$ and $Z0 = c$, we have by definition (2), $e \in S_0^{c,C} y$. As $y \in P_T$ we have by (a, d) that $S_0 y = s_0$, and in particular $S_0^{c,C} y = s_0^{c,C}$. So $e \in s_0^{c,C}$. This argument proves (ℓ) $D_1 \subseteq s_0^{c,C}$.

Now define $\langle e,s \rangle \in Q . \equiv . e \in D_1 \wedge s \in T \wedge P_e \cap P_s \neq 0$. Clearly $Q \subseteq D_1 \times T$. Suppose $e \in D_1$. By (d, f) we have a $s \in T$, such that $P_e \cap P_s \notin \mathcal{O}_1$. This argument shows that Q projects onto D_1. Similarly, (c, g) shows that Q projects onto T. Thus, (m) $Q \subseteq D_1 \times T$, Q onto D_1, Q onto T. Assume $\langle e,s \rangle \in Q$, and $d \in D_1$. Then $e \in D_1$, $s \in T$, and there is a $v \in P_e \cap P_s$. By (d, f) we have $P_T \cap P_d \notin \mathcal{O}_1$, and hence there is a $y \in P_T \cap P_d$, $v < y$. Using (g, i) this yields $\{Zt; v \leq t < y\} = D_0$. Using (h) we have $Zv = e$ and $Zy = d$. Hence, as Z is a run of H, $\mathcal{K}$, it follows by definition (2), that $d \in s_v^{e,D_0} y$. As $v \in P_s$ and $v < y \in P_T$, we have by (b, d), $S_v y = s$. So $d \in s^{e,D_0}$. We have thus shown that $d \in s^{e,D_0}$, for any $d \in D_1$ and any $< e,s > \in Q$. I.e.,

(n) $< e,s > \in Q . \supset . D_1 \subseteq s^{e,D_0}$.

From $\Omega_{s_0,T}$ and $(\exists Z)\Gamma(X,Z)$ we now have found c, C, D_0, D_1, Q such that (e, k, ℓ, m, n). These are just the requirements on the right side of (5). Hence $B[s_0,T]$. It remains to show that from $\Omega_{s_0,T}$ and $B[s_0,T]$ we can find an X-run Z of Γ.

As $\Omega_{s_0,T}$, we still have the P_s and the cofinal-closed P_T which satisfy (a, b, c, d) in the first paragraph of this proof. As $B[s_0,T]$, we have by definition (5), c, C, D_0, D_1 and Q such that $E[c]$, (o) $e \in D_1 \supset \mathcal{K}[D_0,e]$, $\mathcal{L}[D_1]$, $D_1 \subseteq s_0^{c,C}$, (p) $Q \subseteq D_1 \times T$, Q onto D_1 and T, and (q) $< e,s > \in Q \supset D_1 \subseteq s^{e,D_0}$. We are to show that there is an X-run Z of Γ. Of course we will make up Z such that $Z0 = c$, $\sup_1 Z = D_1$ so that the initial and terminal-condition hold. Furthermore D_0 will become $\sup_0 Z$, and Z will be made exact C. This goes as follows.

Suppose $s,r \in T$ and $v \in P_s \cap P_r$. By (c) there is a $y \in P_r$ such that $v < y$. By (b) it follows that $S_v y = s$ and $S_v y = r$, hence $s = r$. This argument shows, (r) $s \neq r \in T \supset P_s \cap P_r = 0$. We now use the fact that $\mathcal{P}\omega_1/\mathcal{I}_1$ has no atom (see Section 5). Because of (c) and because Q is onto T we therefore can split P_s into arbitrarily (finite) many non-null parts. So there exist $P_{e,s}$ such that $P_s = \bigcup_{e,s \in Q} P_{e,s}$, and (c') $< e,s > \in Q \supset P_{e,s} \notin \mathcal{O}_1$, and $< e,s > \neq < d,s > \supset P_{e,s} \cap P_{d,s}=0$. By (r, a, b, d) we now have (r') $< e,s > \neq < d,r > \supset$ $\supset P_{e,s} \cap P_{d,r}=0$, and (a') $< e,s > \in Q \wedge P_{e,s}y \supset S_0y=s_0$, and (b') $< e,s > \in Q \wedge v<y \wedge P_{e,s}v \wedge Py \supset S_vy=s$, and (d') $P=\bigcup_{e,s \in Q} P_{e,s}$ is cofinal-closed. By (r', d') each $v \in P$ belongs to exactly one of the $P_{e,s}$. So we have e_v, s_v such that (s) $Pv \supset < e_v, s_v > \in Q \wedge P_{e_v,s_v}v$, and (t) $< e,s > \in Q \wedge P_{e,s}v \supset e=e_v$. Because of (s, p) we have (u) $Pv \supset e_v \in D_1$.

Let $v \in P$. By (s, u) we have $< e_v, s_v > \in Q$, $P_{e_v,s_v}v$, and $e_v \in D_1$. Because $D_1 \subseteq S_0^{c,C}$ and (a') this implies $e_v \in S_0^{c,C}v$. Hence, by definition (2), there is an $X[0,v)$-run $Z[0,v]$ of $H, \mathcal{K}$, such that $Z0=c$ and $Zv=e_v$. Suppose $v<y$ and $v,y \in P$. By (s, u) we have $< e_v, s_v > \in Q$, $P_{e_v,s_v}v$, and $e_y \in D_1$. Therefore by (q, b'), $e_y \in S_v^{ev,D_0}y$. Hence, by definition (2), there is an $X[v,y)$-run $Z[v,y]$ of $H, \mathcal{K}$, such that $Zv=e_v$, $Zy=e_y$ and $\{Zt; v \leq t < y\}=D_0$. This argument shows the existence of the following runs:

	z the first member of P ;	$v<y$ any consecutive members of P
(*)	$Z[0,z]$ an $X[0,z)$-run of $H, \mathcal{K}$;	$Z0=c$, $Zz=e_z$
	$Z[v,y]$ an $X[v,y)$-run of $H, \mathcal{K}$;	$Zv=e_v$, $Zy=e_y$, D_0-exact

Note that for consecutive members $v<y<w$ of P, the runs $Z[v,y]$ and $Z[y,w]$ have equal states e_y at y, and similarly e_z occurs as state Zz in both runs $Z[0,z]$ and $Z[z,\cdot]$. Hence,

these partial runs can be spliced to make up a run Z. As the splicing places form the cofinal-closed set P, the run Z is a full ω_1-sequence. The segments of Z have the properties (*). It remains to show that Z is an X-run of Γ.

Because of (*) it is clear that Z satisfies the transition-conditions H and $\mathcal{K}$ at every place $v \notin P'$. Suppose now $v \in P'$. From (*) one sees $\sup_0^v Z = D_0$. As P is closed we have $v \in P$, and therefore by (*) and (u), $Zv = e_v \in D_1$. Therefore by (o), $\mathcal{K}[D_0, Zv]$. Thus, the transition-condition $\mathcal{K}$ is satisfied also at $v \in P'$. As E[c] and Z0=c, Z also satisfies the initial-conditions. It remains only to show that Z satisfies the terminal condition $\mathcal{L}[\sup_1 Z]$ at ω_1. Because $\mathcal{L}[D_1]$, this is accomplished if we show $\sup_1 Z = D_1$. This is proved in the remaining two paragraphs.

Suppose first that $e \in D_1$. By (p) there is an s, such that $< e,s > \in Q$. Let $y \in P_{e,s}$. By (d', u) and (*) we see $e = e_y = Zy$. This shows, $P_{e,s} \subseteq \{t; Zt = e\}$. Hence by (c'), $\{t; Zt = e\} \notin \mathcal{O}_1$, i.e., $e \in \sup_1 Z$. This argument shows $D_1 \subseteq \sup_1 Z$.

Suppose now $e \in \sup_1 Z$. Then there is a $U \notin \mathcal{O}_1$, such that $Ut \supset Zt = e$. As P is cofinal closed we have $U \cap P \notin \mathcal{O}_1$. Hence, there is a $v \in U \cap P$. So $Zv = e$, and by (*), $Zv = e_v$. By (d') there is a $< d,s > \in Q$ such that $v \in P_{d,s}$. By (t), $d = e_v$. Thus, $e = Zv = e_v = d$. So $< e,s > \in Q$, and by (p), $e \in D_1$. This argument shows $\sup_1 Z \subseteq D_1$. Q.E.D.

We have used AC_ω^ω in the form "ω_1 is not ω_0-limit" (Lemma 6.3), and we have used $AC_{\omega_1}^\gamma$, $\gamma = 2^\omega$, first to select the parts Z[v,y] to be spliced, and second to show that $\mathcal{J}_1$ admits no atoms. It seems clear how one could do with a somewhat simpler B, if $\mathcal{J}_1$ is prime. We have not investigated whether our proof can be fixed to work without assumptions on $\mathcal{J}_1$. The idea is to work with variations of Ω and B, as to take care of the various possibilities concerning atoms, within the proof of Lemma 6.3. From Lemmas 6.2 and 6.3 we see the following formulas.

$$(6) \quad (\exists Z)\Gamma(X,Z) .\equiv. \bigvee_{\substack{s_0,T \\ B[s_0,T]}} \Omega_{s_0,T} \qquad \sim(\exists Z)\Gamma(X,Z) .\equiv. \bigvee_{\substack{s_0,T \\ \sim B[s_0,T]}} \Omega_{s_0,T}$$

The second formula is the rough form of the complementation lemma at ω_1. It remains to use the subset-recursion for countable ordinals, to express $\Omega_{s_0,T}$ by a transition-system of form (1).

<u>Lemma</u> 6.4: One can construct matrices $E'[Y0]$, $H'[Xt,Yt,Yt']$, and $\mathcal{K}'[\sup_0^x Z, Zx]$, and, $\mathcal{L}_{s_0,T}[\sup_1 Y]$, such that

$$\Omega_{s_0,T} \equiv (\exists Y).E'(0) \wedge (\forall t)H'(t) \wedge (\forall x)\mathcal{K}'(x) \wedge \mathcal{L}_{s_0,T}[\sup_1 Y]$$

<u>Proof</u>: Let $Y0=e$, $Yt'=F[Xt,Yt]$, $Yx=\mathcal{J}[\sup_0^x Y]$ be the subset-recursion of Lemma 4.4. Note that Y has a component Y^0 (denoted S_0), of which we know $Y^0t=S_0t$, for all $t<\omega_1$, and arbitrary input. Let F^0 and $\mathcal{J}^0$ be the corresponding component

of F and $\mathcal{S}$ respectively. Many copies of the above recursion will make up the essential part of the transition-conditions H', $\mathcal{K}'$. Note that a copy, started at $Yv=e$, will yield $Y^0t=S_vt$, for all t. Just as in the proof of Lemma 3.5, we make up the following transition-system. s ranges over T, and Pt stands for $\bigvee_s P_st$:

$$\left.\begin{array}{lll} Y_00=e & & \tilde{Q}_s0 \end{array}\right\} E'$$

$$\left.\begin{array}{rrl} & Y_0t'=F[Xt,Y_0t] & Q_st' \equiv [Q_st \vee P_st] \\ \tilde{Q}_st \wedge P_st \supset Y_st=e & Q_st' \supset Y_st'=F[Xt,Y_st] & \\ P_st \supset \overline{Y}_st=e & \tilde{P}t' \wedge Q_st' \supset \overline{Y}_st'=F[Xt,\overline{Y}_st] & Pt' \wedge Q_st' \supset F^0[Xt,\overline{Y}_st]=Y_s^0t' \end{array}\right\} H'$$

$$\left.\begin{array}{rr} Y_0x=\mathcal{S}[\sup_0^x Y_0] & Q_sx \equiv (\exists_0^x t)Q_st \\ Q_sx \supset Y_sx=\mathcal{S}[\sup_0^x Y_s] & \\ (\forall_0^x t)\tilde{P}t \wedge \tilde{P}x \wedge Q_sx \supset \overline{Y}_sx=\mathcal{S}[\sup_0^x \overline{Y}_s] & (\forall_0^x t)\tilde{P}t \wedge Px \wedge Q_sx \supset \mathcal{S}^0[\sup_0^x \overline{Y}_s]=Y_s^0x \end{array}\right\} \mathcal{K}'$$

$$\left.\bigwedge_{s\in T}(\exists_1 y)P_sy \wedge (\forall_1 y)[\bigvee_{s\in T} P_sy \wedge Y_0^0y=s_0 \wedge \bigwedge_{s\in T} Y_s^0y=s]\right\} \mathcal{L}_{s_0,T}$$

Suppose now $\Omega_{s_0,T}$. By (4) there are sets P_s, $s\in T$ such that $P_s \notin \mathcal{O}_1$, $\cup P_s \in \mathcal{J}_1$, $P_sy \supset S_0y=s_0$, and $v<y \wedge P_sv \wedge P_ry \supset S_vy=s$. When we delete from E', H', $\mathcal{K}'$ the clauses $Pt' \wedge Q_st' \supset F^0[Xt,\overline{Y}_st]= Y_s^0t'$ and $(\forall_0^x t)\tilde{P}t \wedge Px \wedge Q_sx \supset \mathcal{S}^0[\sup_0^x \overline{Y}_s]=Y_s^0x$, the remaining formulas define recursively, from the P_s, sequences Y_0, Q_s, Y_s, $\overline{Y}_s$. (Incidentally, if $P_sv \wedge (\forall t)_v^z \cdot \tilde{P}t \wedge P_rz$ for some $r \neq s$, then $\overline{Y}_s$ is not determined at z and thus remains undetermined up to the next y such that P_sy, where $\overline{Y}_s$ is reset to e. As the values of $\overline{Y}_s$ will not be needed at these places, we may choose $\overline{Y}_s=Y_s$ in these intervals.) Let u_s be the first member of P_s. Note that Q_s takes the value F up to u_s', and remains T from there on. Y_s is started at u_s in the state e, and from there on is defined by the subset-recursion. Hence, $Y_s^0t=S_{u_s}t$ for all $t \geq u_s$. As $u_s \in P_s$ we therefore have, $u_s<y \in P_r \supset Y_s^0y=s$. In a similar

fashion we can show that the sequences $Y_0, Q_s, Y_s, \overline{Y}_s$ satisfy $Pt' \wedge Q_s t' \supset F^0[Xt, \overline{Y}_s t] = Y_s^0 t'$ and $(\forall_0^x t)\tilde{P}t \wedge Px \wedge Q_s x \supset \mathcal{J}^0[\sup_0^x \overline{Y}_s] = Y_s^0 x$. Y_0 is defined from 0, by the subset-recursion. Hence, $Y_0^0 t = S_0 t$ for all t. Therefore, $P_s y \supset Y_0^0 y = s_0$. So, with exception of the finitely many u_s, $y \in \bigcup P_s \supset . Y_0^0 y = s_0 \wedge Y_s^0 y = s$. As $P_s \notin \mathcal{O}_1$ and $\bigcup P_s \in \mathcal{J}_1$, we therefore see that $P_s, Q_s, Y_0, Y_s, \overline{Y}_s$ is a run of $E', H', \mathcal{K}'$, which satisfies the terminal condition $\mathcal{L}_{s_0,T}$. It remains to show that if such a run exists, then $\Omega_{s_0,T}$.

Suppose then that $P_s, Q_s, Y_0, Y_s, \overline{Y}_s$ is a run of our system $E', H', \mathcal{K}', \mathcal{L}_{s_0,T}$. Let $\overline{P}_s$ be the intersection of P_s with the finitely many sets $\{y; Y_0^0 y = s_0\}$ and $\{y; Y_r^0 y = r\}$, $r \in T$. Because $\mathcal{L}_{s_0,T}$ and $\mathcal{J}_1$ is a filter, we still have $(\forall_1 y) \bigvee_{s \in T} \overline{P}_s y$ and $(\exists_1 y) \overline{P}_s y$, for $s \in T$. To see $\Omega_{s_0,T}$, it therefore suffices by (4), to show $s \in T \wedge \overline{P}_s y \supset S_0 y = s_0$, and $s, r \in T \wedge v < y \wedge \overline{P}_s v \wedge \overline{P}_r y \supset S_v y = s$. By the definition of the $\overline{P}_s$, this will be accomplished if we show (a) $S_0 y = Y_0^0 y$, and (b) $v < y \wedge P_s v \wedge P_r y \supset S_v y = Y_s^0 y$. Now, inspecting our system $E', H', \mathcal{K}'$ shows that Y_0 is given by the subset-recursion, starting at 0. This proves (a). Suppose now $v < y \wedge P_s v \wedge P_r y$. Inspecting our system shows that at time v, $\overline{Y}_s$ is reset to e. From there on $\overline{Y}_s$ is subjected to the subset-recursion until a $z \leq y$, $z \in \bigcup P_s$ is met. At this time z, $\overline{Y}_s$ feeds into Y_s, which is subjected to the subset-recursion. Hence we have $\overline{Y}_s^0 t = S_v t$ for $t < z$, and $Y_s^0 t = S_v t$ for $z \leq t$. It follows that $Y_s^0 y = S_v y$, which proves (b). This ends the proof of our lemma; $\mathcal{L}_{s_0,T}$ can be put into the required $\sup_1$-form. Q.E.D.

<u>Lemma</u> 6.5 (<u>complementation at</u> ω_1): To every transition-system $\Gamma(X,Z)$ of form (1), one can construct a system $\Gamma'(X,Y)$ of the

same form, such that $\sim(\exists Z)\Gamma(X,Z) .\equiv. (\exists Y)\Gamma'(X,Y)$.

Proof: By (6) and Lemma 6.4 we have $\sim(\exists Z)\Gamma(X,Z)$ just in case $\bigvee_{s_0,T}(\exists Y).E'(0) \wedge (\forall t)H'(t) \wedge (\forall x)\mathcal{X}'(x) \wedge \mathcal{L}_{s_0,T}[\sup_1 Y]$, the disjunction ranging over $s_0,T \in \tilde{B}$. This disjunction can clearly be moved in front of $\mathcal{L}_{s_0,T}$. It remains only to put $\bigvee_{s_0,T}\mathcal{L}_{s_0,T}[\sup_1 Y]$ into form $\mathcal{L}'[\sup_1 Y]$. Q.E.D.

From the complementation lemma, and the prenex-form lemma 1.5 we obtain, in the usual fashion, the following:

Theorem 6.6 (definability in MT of ω_1): To every formula $\Sigma(X)$ of $MT[\omega_1,<]$ one can construct a transition-system Γ of form (1), such that for any ω_1-sequence X, $\Sigma(X)$ holds in $[\omega_1,<]$ just in case Γ accepts X.

The prenex form lemma 1.5 will at first give a system Γ_0, which is like (1), except that the terminal-condition is a $\sup_0$-condition, $\mathcal{L}_0[\sup_0 Y]$. Inspecting the proof of Lemma 1.5 shows that $\mathcal{L}_0[\sup_1 Y]$ can be put in place. Alternatively: a $\sup_0$-condition can be put into $\sup_1$-form, by using remark 5.3.

Once the theory of regular events or strings of length $<\omega_1$ is worked out, this theorem will yield rather concrete information on definability in $[\omega_1,<]$, by MT-formulas. We will now restate the input-free case.

<u>Theorem</u> 6.7 (<u>the normal form of sentences</u>): To every MT-sentence Σ in $<$, one can construct matrices $E[Z0]$, $H[Zt,Zt']$, $\mathcal{K}[\sup_0^x Z, Zx]$, and $\mathcal{L}[\sup_1^x Z]$, such that Σ holds in $[\omega_1,<]$ just in case $(\exists Z).E(0) \wedge (\forall t)H(t) \wedge (\forall x)\mathcal{K}(x) \wedge \mathcal{L}(\omega_1)$.

In other words, deciding truth of sentences in $MT[\omega_1,<]$ comes to deciding whether input-free systems of the following form admit a run.

$$(1') \quad \Gamma(Z) \quad : \quad E[Z0] \wedge (\forall t)H[Zt,Zt'] \wedge (\forall x)\mathcal{K}[\sup_0^x Z, Zx] \wedge \mathcal{L}[\sup_1 Z]$$

To find a decision-method for this problem, we copy the ideas presented in Remark 3.9.

<u>Remark</u> 6.8: Given the input-free system Γ in (1'), one can construct s, such that $\Omega_{s,\{s\}}$.

<u>Proof</u>: As the system is input-free we clearly have (a) $S_0 t = S_u(u+t)$. Using Lemma 4.7, and the recursions in Lemma 4.10, we can find a number p, and a state s, such that $S_0(\omega^p t) = s$, for all t. Using (a) this yields $\omega^p t \approx 0(\omega^p t + \omega^p)$, and therefore $\omega^p t \approx 0(-\omega_1)$. But the $\omega^p t$ form a cofinal-closed subset of ω_1. Hence, $(\forall_1 y) y \approx 0(-\omega_1)$ and $(\forall_1 y) S_0 y = s$. Letting $W=\{0\}$, we therefore see by definition (3), that $\Omega_{s,\{s\}}$. Q.E.D.

<u>Theorem</u> 6.9: The monadic second order theory of $[\omega_1,<]$ is decidable, if AC_ω^γ whereby γ is the power of the continuum.

Proof: Given the sentence Σ, we use the procedures indicated in the long proof (it extends through both Sections 4 and 6) of theorem 6.7, to construct an input-free system (1'), such that Σ holds in $[\omega_1,<]$ if and only if $(\exists Z)\Gamma(Z)$. Next we construct the s of Remark 6.8 (by using the decision-method in Section 4), so $\Omega_{s,\{s\}}$. By Lemma 6.3, we now have $(\exists Z)\Gamma(Z)$ holds just in case $B[s,\{s\}]$. So the truth-value of Σ can actually be found by evaluating B, using its definition (5). Q.E.D.

Actually we have shown that our decision-method works under the assumptions: ω_1 is not an ω_0-limit, and $\mathcal{P}\omega_1/\mathcal{I}_1$ has no atoms. In Section 5 we showed that these assumptions follow from AC_ω^γ. It would be interesting to have a decision method for $MT[\omega_1,<]$, which works in case ω_1 should be an ω_0-limit. The methods of this section can easily be adapted to the case: ω_1 not ω_0-limit, and $\mathcal{I}_1$ is prime. In fact this turns out much simpler, as now $(\exists_1 y)\equiv(\forall_1 y)$. Finally we recall that all our proofs make use of accessibility properties, rather than cardinality, excepting the Ulam-argument in Section 5. The results are therefore better stated, by replacing ω_1 by $\varkappa_1$. It remains to show a decision-method for $MT[\varkappa_1,<]$, which works no matter whether $\mathcal{I}_1$ does or does not admit atoms.

BIBLIOGRAPHY

1. J.R. Büchi, Weak second order arithmetic and finite automata, Z. Math. Logik und Grundl. Math. 6 (1960), 66-92.

2. __________, On a decision method in restricted second order arithmetic, Proc. 1960 Int. Cong. for Logic, Method and Philos. of Sci., Stanford Univ. Press, Stanford, Calif., 1962.

3. __________, Decision methods in the theory of ordinals, Bull. Am. Math. Soc. 71 (1965), 767-770.

4. __________, Transfinite automata recursions and weak second order theory of ordinals, Proc. 1964 Int. Cong. for Logic, Method and Philos. of Sci., North-Holland Publishing Co., Amsterdam, 1965.

5. Büchi and Landweber, Solving sequential conditions by finite-state strategies, Trans. Am. Math. Soc. 138 (1969), 295-311.

6. ____________________, Definability in the monadic second-order theory of successor, Journ. Sym. Logic 34 (1969), 166-170.

7. A. Church, Logic arithmetic and automata, Proc. Int. Congr. Math 1962, Almquist and Wiksells, Uppsala, 1963.

8. J. Doner, Tree acceptors and some of their applications, Journ. Computer and System Sci. 4 (1970), 406-451.

9. A. Ehrenfeucht, Application of games to the completeness problem for formalized theories, Fund. Math. 49 (1960-61), 129-141.

10. C.C Elgot, Decision problems of finite automata design and related arithmetics, Trans. Am. Math. Soc. 98 (1961), 21-51.

11. Feferman and Vaught, The first order properties of products of algebraic systems, Fund. Math. 47 (1959), 57-103.

12. H. Läuchli, A decision procedure for the weak second order theory of linear order, Contributions to mathematical logic, Proceeding of the Logic Colloquium, Hannover, 1966, North-Holland, Amsterdam, 1968.

13. R. McNaughton, Testing and generating infinite sequences by a finite automaton, Information and Control 9 (1966), 521-530.

14. M.O. Rabin, Decidability of second-order theories and automata on infinite trees, Trans. Am. Math. Soc. 141 (1969), 1-35.

15. D. Siefkes, Büchi's monadic second order successor arithmetic, Lecture Notes in Mathematics, vol. 120, Springer-Verlag, Berlin, 1970.

16. Th. Skolem, Untersuchungen über die Axiome des Klassenkalküls und über Produktations-und Summationsprobleme, welche gewisse Klassen von Aussagen betreffen, Skrifter, Videnskabsakademiet i Kristiania, no. 3, 1919.

17. S. Ulam, Zur Masstheorie in der allgemeinen Mengenlehre, Fu. Ma. 16 (1930), 140-150.

18. L. Löwenheim, Über Möglichkeiten im Relativkalkül, Fund. Math. 76 (1915), 447-470

AXIOMATIZATION OF THE MONADIC SECOND ORDER THEORY OF ω_1

J. Richard Büchi, Dirk Siefkes

For any ordinal α, let $MT[\alpha]$ be the Monadic second order Theory of α, i.e. the theory where quantification over both, ordinals less than α and subsets of α, is allowed and the order relation is the only primitive. Let $MT[co]$ be the Monadic second order Theory of Countable Ordinals, i.e. the intersection of all $MT[\alpha]$ for $\alpha<\omega_1$. $MT[\alpha]$ for $\alpha<\omega_1$ and $MT[co]$ are decidable by Büchi [Bü 4]. The decidability of $MT[\omega_1]$ is proved in the preceding paper [Bü], where also detailed proofs for the results of [Bü 4] are given. In this paper we characterize $MT[co]$ and $MT[\alpha]$ for $\alpha\leqslant\omega_1$ by axiom systems. A further paper [BS 1] on the subject is planned: There all consistent and complete extensions of $MT[co]$ (which include $MT[\alpha]$ for $\alpha<\omega_1$) will be characterized by axiom systems; it will be proved that the Boolean algebra of sentences of $MT[co]$ has an ordered basis of type $1+2\cdot\eta$.

In the <u>standard</u> <u>interpretation</u> of an MS-formula the set quantifiers range over all subsets of the domain; in general, in an <u>interpretation</u> the set quantifiers may be restricted to a certain family of subsets of the domain. A set $\mathfrak{A}$ of sentences is a <u>(standard) axiom system</u> for the MS-theory $\mathcal{T}$ iff $\mathfrak{A}$ is recursive and exactly the true sentences of $\mathcal{T}$ are the (standard) consequences of $\mathfrak{A}$, i.e. hold in all (standard) models of $\mathfrak{A}$. Caution: What we call "standard axiom system" here, is called simply "axiom system" in [Bü]. Actually, for $\alpha<\omega^\omega$, the standard axiom systems for $MT[\alpha]$ which are easily drawn from [Bü], are axiom systems; for $\alpha=\omega$ this was shown in Siefkes [Si 1]. For $\alpha\geqslant\omega^\omega$, there are certain consequences of the axiom of choice which are independent from the standard axioms for $MT[\alpha]$, i.e. which need not hold in (general) models of the standard axioms. Thus we get the axiom systems for MT[co] and for MT[α] where $\omega^\omega \leq \alpha \leq \omega_1$, by extending the standard axiom systems of [Bü] by these choice axioms. It is open whether the (general) consequences of the standard axioms alone yield decidable theories.

The results of [Bü] and of this paper seem to furnish the first known examples of theories whose decidability depends on the underlying set theory. See also the paper of Litman [Li] where the situation at ω_1 is clarified (cf. the beginning of section 8). Thus, to look for a "nice" axiom system for a decidable theory is more than a matter of esthétics: an axiom system shows precisely how much set theory is involved in the decision procedure. When this paper was finished we learned from the paper [Sh] of Shelah that at ω_2 the situation becomes even more drastic than at ω_1 : In MT[ω_2] statements are expressible which are independent from ZFC . The rôle of the axiom of choice in MT[co] , however, and even more in MT[ω^ω] , is surely unexpected.

Here and throughout the paper the statement

(A) AC is used in the decidability proof for the theory $\mathcal{T}$

has the following meaning: $\mathcal{T}$ is described semantically, i.e. $\mathcal{T}$ is the set of sentences which are true in a certain class $\mathcal{K}$ of structures. $\mathcal{K}$ is given through the models of ZFC (Zermelo-Fraenkel) set theory plus the axiom of choice). There are models of ZF where AC fails and the corresponding class $\mathcal{K}'$ of structures yields a different theory: $\mathcal{T}(\mathcal{K}') \neq \mathcal{T}(\mathcal{K}) = \mathcal{T}$. Thus AC is needed to determine $\mathcal{T}$ and therefore the decision procedure for $\mathcal{T}$.

The statement (A) could also be used with the following different meaning: AC is needed to prove that the set M of Gödel numbers of the true sentences of $\mathcal{T}$ is recursive. If M is defined by a formula of arithmetic, then "M is recursive" is an arithmetical statement. Any arithmetical statement which is provable within ZFC , can be proved without AC . Therefore, if a decidable theory $\mathcal{T}$ is arithmetically describable, especially if $\mathcal{T}$ is axiomatizable within a finitary derivation system, then the decidability of $\mathcal{T}$ can be proved without AC . The hypothesis on the describability of $\mathcal{T}$ cannot be dropped: $\mathcal{T}$ may well be decidable, and therefore M recursive and thus definable in arithmetic; but it might be that one needs AC for the proof that M is the set of Gödel numbers of the true sentences of $\mathcal{T}$. It seems that no case of this use of AC in a decidability proof is known which is not of the type considered in the previous paragraph.

Let us call a theory $\mathcal{T}$ *elementary* iff exactly the true sentences of $\mathcal{T}$ are derivable within a finitary logic calculus. $\mathcal{T}$ is a *first-order* theory on the other hand iff all its variables are individual variables, i.e. range over the elements of possible domains. It might be called Skolem's thesis (see e.g. Skolem [Sk 1,2,3]) that any theory becomes elementary if the hidden primitives are made explicit, i.e. if there is no intended meaning in the symbols which is not caught by the axioms, This would show that the only true meaning of the statement (A) is the one discussed first, which is treated in this paper.

The foregoing considerations will be made more precise in section 1 . The contents of this paper can best be seen from the detailed table of contents given in the beginning.

Acknowledgment: The contents of this paper were first presented in a seminar of the second author at Purdue University in spring 1971. We are grateful for helpful remarks of John Doner and Steve Klein in this seminar. We thank Charles Zaiontz for careful reading of an earlier version of this paper which resulted in valuable criticism. Especially he suggested that an originally given derivation of the splicing axiom from the standard axioms was erraneous; only then we proved the independence of the splicing axiom. This independence proof uses unpublished work of Litman [Li] . We are indebted to Ami Litman for his kind permission to present his proof, to Gadi Moran who translated and explaned us parts of [Li] , and to Ulrich Felgner who brought to our attention the work of Church, Hájek, and Specker quoted in section 8 . - The results of this paper were reported at the Tarski Symposium in Berkeley, June 1971.

1. Axiomatization of monadic second order theories

We want to axiomatize the monadic second order theories of ordinals which are shown to be decidable in the preceding paper [Bü] . Thus the monadic second order language involved, call it $\mathcal{L}$, uses the following symbols:

individual variables $t,u,\dots,z,$

set variables $P,Q,\dots,Z,$

primitives $\in , = , <,$

logical connectives $\wedge , \vee , \neg , \longrightarrow , \longleftrightarrow , \forall , \exists ,$

brackets and dots.

Other symbols will be introduced as necessary. Prime formulae are of the form

$$x \in Z \ , \ x = y \ , \ X = Y \ , \ x < y \ ,$$

arbitrary formulae are built up from these in the usual way. For examples see section 1 of [Bü] . We will overtake most of the notation and terminology of that paper.

The structures we want to describe are of the form

$$[\alpha] = \langle \alpha , < \rangle \ ,$$

where α is an ordinal and $<$ is the natural order on α . Let ω_1 be the first uncountable ordinal. As in [Bü] let $MT[\mathcal{K}]$, for a class $\mathcal{K}$ of structures, be the set of all sentences of $\mathcal{L}$ which hold in all structures of $\mathcal{K}$: the monadic second order theory of $\mathcal{K}$. Then the theories which we want to axiomatize are $MT[\alpha] =_{df} MT[\{[\alpha]\}]$ for $\alpha < \omega_1$, and $MT[co] =_{df} MT[\{[\alpha];\ \alpha < \omega_1\}]$. It should be recalled that in interpreting formulae in the model $[\alpha]$ the range of the set variables is $\mathcal{P}(\alpha)$, the full power set of α , and the symbol $\in$ is interpreted as the membership relation. Following the terminology of Henkin [He 1] we will call this the <u>standard interpretation</u> of monadic second order formulae.

To see that we cannot restrict ourselves to standard interpretations, we consider the elementary theory of ordinals, ET[o] . Let ω^* be the reverse order of ω , i.e. the order type of the negative integers. Then according to Mostowski and Tarski [MT 1] , $\omega + \omega^*$ is a model of ET[o] . In fact, enlarge ET[o] by an infinite set of sentences stating that

(i) there are infinitely many elements,

(ii) there is a last element,

(iii) there are no limit numbers.

Then this extension of ET[o] , call it ET$[\omega + \omega^*]$, is consistent. Indeed, any finite set of sentences of ET$[\omega + \omega^*]$ has a (finite) model, therefore ET$[\omega + \omega^*]$ has a model by the compactness theorem. It is easily seen that $\omega + \omega^*$ is a model of ET$[\omega + \omega^*]$. Actually it is the only one up to elementary equivalence, since by Mostowski-Tarski [MT 2] ET$[\omega + \omega^*]$ is complete. On the other hand, $\omega + \omega^*$ is not a model of MT[co] under standard interpretation, since its domain is not well-ordered: the subset ω^* violates the minimum principle. Even more, if we add the above conditions (i) - (iii) to MT[co] , then the resulting extension is inconsistent: it has no model under standard interpretation.

Thus we have to drop the restriction to standard interpretations, if we want to keep models of the elementary theory for the monadic second order case. Let $\mathcal{P}_{\text{uper}}(\omega + \omega^*)$ consist of those subsets of $\omega + \omega^*$ which, with the exception of finitely many elements, are of the form $M \cup M^*$ where M is a periodic subset of ω and M^* is the corresponding reversed subset of ω^* . Then we will show in [BS 1] that $\omega + \omega^*$ becomes a model of MT[co] if we restrict the range of the set variables to $\mathcal{P}_{\text{uper}}(\omega + \omega^*)$. In general, we will have to consider general models in the sense of Henkin [He 1] , where the range of the set variables is a subset of the power set of the domain. Actually we will show in [BS 1] that there are $2^{\aleph_0}$ possible ways to extend the ultimately periodic subsets of ω to ultimately periodic subsets of $\omega + \omega^*$, as we did it above. That means, there are $2^{\aleph_0}$ extensions of MT[co] which have general models over the domain $\omega + \omega^*$, but which do not have a standard model. Thus we would loose most of the richness of MT[co] if we restrict ourselves to standard interpretation.

For a moment let us regard monadic second order theories as two-sorted elementary theories. We will recall some basic facts on model theory for (many-sorted) elementary theories for the case of the language $\mathcal{L}$ and a given class of models. A <u>structure</u>, which can be used to interprete formulas of $\mathcal{L}$, is now of the form

$$\mathcal{D} = \langle D_0 , D_1 ; E , = , O \rangle ,$$

where D_0 and D_1 are any two sets, E is a relation between elements of D_0 and elements of D_1 (i.e. a subset of $D_0 \times D_1$) , = is the equality on D_0 and D_1 , and O is a binary relation on D_1 . A structure $\mathcal{D}$ is a <u>model</u> of a sentence Υ of $\mathcal{L}$,

$$\Upsilon \succ\!\!- \mathcal{D} ,$$

in case Υ becomes true if the individual variables range over D_0 and the set variables range over D_1 , and if $\in$ and $<$ are interpreted as E and O resp. (Thus = is always interpreted as the equality.) For the next few pages let $\mathcal{M}$ be the class of all structures, and $\mathcal{T}$ the set of all sentences of $\mathcal{L}$. The model relation constitutes a Galois correspondence (S,T) between sets of sentences of $\mathcal{L}$ and classes of structures in $\mathcal{M}$ (see e.g. Birkhoff [Bi] , V.7+8), as follows. For sets $\mathcal{A}, \mathcal{B} \subseteq \mathcal{T}$ and for classes $\mathcal{G}, \mathcal{H} \subseteq \mathcal{M}$ we define:

(i) $S[\mathcal{A}] =_{df} \bigcap_{\Upsilon \in \mathcal{A}} \{\mathcal{D} \in \mathcal{M} ;\ \Upsilon \succ\!\!- \mathcal{D}\}$,

the <u>class of models of</u> $\mathcal{A}$,

(ii) $T[\mathcal{G}] =_{df} \bigcap_{\mathcal{M} \in \mathcal{G}} \{\Upsilon \in \mathcal{T} ;\ \Upsilon \succ\!\!- \mathcal{D}\}$,

the <u>theory of</u> $\mathcal{G}$.

The operators S and T have the following properties:

(iii) $\mathcal{A} \subseteq \mathcal{B}$ implies $S[\mathcal{B}] \subseteq S[\mathcal{A}]$,

(iv) $\mathcal{G} \subseteq \mathcal{K}$ implies $T[\mathcal{K}] \subseteq T[\mathcal{G}]$.

Further, ST and TS are closure operators on $\mathcal{M}$ and $\mathcal{T}$, resp. I.e.

(v) $\mathcal{A} \subseteq TS[\mathcal{A}]$,

(vi) $\mathcal{A} \subseteq \mathcal{B}$ implies $TS[\mathcal{A}] \subseteq TS[\mathcal{B}]$,

(vii) $TSTS[\mathcal{A}] = TS[\mathcal{A}]$,

and similarly for ST . Actually we have even

(vii*) $STS[\mathcal{A}] = S[\mathcal{A}]$, $TST[\mathcal{G}] = T[\mathcal{G}]$

The operator TS is normally called (logical) consequence, here rather $\mathcal{M}$- <u>consequence</u>; for better reading we will write cl($\mathcal{A}$) instead of $TS[\mathcal{A}]$. The sets closed under this operator are $\mathcal{M}$- <u>theories</u>. A set $\mathcal{A}$ is an $\mathcal{M}$- <u>axiom system</u> for the $\mathcal{M}$- theory $\mathcal{C}$ iff $\mathcal{A}$ is recursive and cl($\mathcal{A}$) = $\mathcal{C}$. (Note that then $\mathcal{A} \subseteq \mathcal{C}$.) It is easy to see that for an $\mathcal{M}$- theory $\mathcal{C}$ and a recursive set $\mathcal{A}$ of sentences the following three conditions are equivalent:

(a) $\mathcal{A}$ is an $\mathcal{M}$- axiom system for $\mathcal{C}$ (i.e. for all sentences $\Upsilon : \Upsilon \in$ cl ($\mathcal{C}$) = $\mathcal{C}$ iff ($\Upsilon \in$ cl ($\mathcal{A}$))

(b) For all sentences Υ and sets of sentences $\mathcal{B}$: $\Upsilon \in$ cl ($\mathcal{B} \cup \mathcal{C}$) iff $\Upsilon \in$ cl ($\mathcal{B} \cup \mathcal{A}$)

(c) For all structures $\mathcal{D} \in \mathcal{M} : \mathcal{D} \in S(\mathcal{C})$ iff $\mathcal{D} \in S(\mathcal{A})$

An $\mathcal{M}$- axiom system $\mathcal{A}$ is $\mathcal{M}$- <u>consistent</u> if it has a model in $\mathcal{M}$; it is $\mathcal{M}$- <u>complete</u> if for any sentence Υ either Υ or $\neg\Upsilon$ is an $\mathcal{M}$- consequence of $\mathcal{A}$, or equivalently: if any two models of $\mathcal{A}$ in $\mathcal{M}$ are elementarily equivalent (satisfy the same sentences); it is $\mathcal{M}$- <u>categorical</u> if any two models of $\mathcal{A}$ in $\mathcal{M}$ are isomorphic. We will use these definitions for theories as well as for axiom systems; we will call an $\mathcal{M}$- complete $\mathcal{M}$- theory <u>complete</u>. By Gödel's completeness theorem for elementary logic $\mathcal{M}$- consequence can be identified with derivability within a derivation system for the first order predicate calculus.

Besides this connection to derivability, everything remains unchanged if we replace $\mathcal{M}$ by a smaller class $\mathcal{N}$. But note that the operator S , but not T , depends on the choice of $\mathcal{N}$. Indeed

$$S_{\mathcal{N}}[A] = S_{\mathcal{M}}[A] \cap \mathcal{N} .$$

Since thus

$$A \subseteq cl_{\mathcal{M}}(A) \subseteq cl_{\mathcal{N}}(A) ,$$

any $\mathcal{M}$- consequence is an $\mathcal{N}$- consequence, any $\mathcal{N}$- theory is an $\mathcal{M}$-theory, any $\mathcal{M}$- axiom system for an $\mathcal{N}$- theory is an $\mathcal{N}$- axiom system.

We call a structure D <u>standard</u> if $D_1 = \mathcal{P}(D_0)$ and E is the membership relation. In case $\mathcal{N}$ is the class of all standard structures, we will write "S*" and "cl*" instead of "$S_{\mathcal{N}}$" and "$cl_{\mathcal{N}}$" resp., and replace "$\mathcal{N}$" by "standard" everywhere else: <u>standard</u> <u>consequence</u>, <u>standard</u> <u>complete</u> and so on. Also we will drop the "$\mathcal{M}$" now, saying "<u>consequence</u>" instead of "$\mathcal{M}$- consequence" etc. It is worth recalling that any consequence is a standard consequence, but not conversely; analogously with "axiom system", "complete", "categorical". Note that the operator S* coincides with the operator MS of section 1 of [Bü] , only that there the power set of the domain is not explicitly mentioned as range for the set variables. The operators T and MT coincide on sets of standard structures; otherwise MT is not defined. To make our notation uniform, we will change the definition of the standard structure $[\alpha]$ for any ordinal α into:

$$[\alpha] =_{df} \langle \alpha , \mathcal{P}(\alpha) ; \epsilon , = , < \rangle .$$

Then we can redefine $MT[\alpha] = T[\{[\alpha]\}]$ and $MT[co] = T[\{[\alpha] ; \alpha < \omega_1\}]$.

From (vii*) above we infer that for any set A of sentences, A is a (standard) theory iff $A = T[\mathcal{K}]$ for some class $\mathcal{K}$ of (standard) structures. Thus MT $[\alpha]$ and MT[co] are standard theories. We note as a special consequence: If a sentence holds in a model of MT[co], then it holds already in a standard model of MT[co], i.e. it holds in a structure $[\alpha]$ for a countable ordinal α.

Let $\mathcal{P}_1$ be any derivation system for the two-sorted first-order predicate calculus with equality, restricted to our language $\mathcal{L}$. Then Gödel's completeness theorem yields for any sentence Ψ and any set A of sentences: $\Psi \in \mathrm{cl}\,(A)$ iff Ψ is derivable from A within $\mathcal{P}_1$. An easy consequence is the compactness theorem: A set of sentences is consistent iff all its finite subsets are consistent. The above example of the theory $\mathrm{ET}[\omega + \omega^*]$, which motivated the introduction of non-standard structures, shows that we cannot have similar results for "standard consequence" instead of "consequence". In fact, let A_0 be the extension of MT[co] by the set of sentences (i)-(iii) in the beginning of this section which defined the theory $\mathrm{ET}[\omega + \omega^*]$. As remarked there, A_0 has no standard model, i.e. it is standard inconsistent. As in the elementary case, however, any finite subset of A_0 has a (finite) standard model, thus is standard consistent. In short: monadic second order logic for standard models is incompact. Therefore there can be no finitary system of rules and axioms which allows to identify derivability and standard consequence. By the way, $\mathrm{cl}\,(A_0)$ is an example of a theory which is not a standard theory: $\mathrm{cl}\,(A_0) \subsetneq \mathrm{cl}^*(A_0) = \mathcal{T}$.

The completeness proof of Henkin [He 1] for the simple theory of types shows that one can go a small step toward standard structures. We call a structure <u>general</u> if D_1 is a subfield of the powerset $\mathcal{P}(D_0)$ which contains all definable subsets of D_0, and E is the membership relation. Here $M \subseteq D_0$ is <u>definable</u> iff there is a formula $\Psi(x)$ with no other free variables such that for all $d \in D_0$:

$$d \in M \quad \text{iff} \qquad \Psi[x/d] \succ\!\!- \langle D_0, \mathcal{P}(D_0) \; ; \; \epsilon \; , \; = \; , \; 0 \rangle$$

where $[x/d]$ means that x is interpreted by d. <u>General</u> <u>consequence</u>

and so on is then defined in the obvious way. Obviously any standard structure is general. We obtain the derivation system $\mathcal{P}_2$ from $\mathcal{P}_1$ by adding the extensionality axiom

(EXT) $$(\forall z)[Xz \longleftrightarrow Yz] \longrightarrow X = Y$$

and the comprehension axiom schema

(COMP) $$(\exists Z)\ (\forall x)[Zx \longleftrightarrow \Upsilon(x)]\ ,\ Z \text{ not in } \Upsilon .$$

(Here we use the usual convention that an axiom is understood as the universal closure of the indicated formula.) Note that the dual of (EXT) , i.e.

$$(\forall Z)[Zx \longleftrightarrow Zy] \longrightarrow x = y\ ,$$

is derivable from (Comp) . The converse statements of these two are derivable already within $\mathcal{P}_1$. Henkin's proof (combined with Henkin [He 2] ; see also Siefkes [Si 1] where a system equivalent to $\mathcal{P}_2$ is used) yields for any sentence Υ and any set $\mathcal{A}$ of sentences: Υ is a general consequence of $\mathcal{A}$ iff Υ is derivable from $\mathcal{A}$ within $\mathcal{P}_2$. In fact this is an easy consequence of Gödel's completeness theorem for first order logic: Υ is derivable from $\mathcal{A}$ within $\mathcal{P}_2$ iff Υ is derivable from $\mathcal{A} \cup \{(EXT)\ ,\ (COMP)\}$ within $\mathcal{P}_1$ iff $\Upsilon \in cl\ (\mathcal{A} \cup \{(EXT)\ ,\ (COMP)\})$. This last statement implies directly that Υ is a general consequence of $\mathcal{A}$. To get the converse of this implication, consider any model $\mathcal{D} = \langle D_0, D_1; E, =, 0\rangle$ of $\mathcal{A} \cup \{(EXT)\ ,\ (COMP)\}$. Define for any $X \in D_1$,

$$f(X) =_{df} \{x \in D_0\ ;\ xEX \text{ holds in } \mathcal{D}\} .$$

Then f is a one-to-one function from D_1 into $\mathcal{P}(D_0)$ such that: xEX iff $x \in f(X)$. Since $\mathcal{D}$ satisfies (COMP) , $\mathcal{D}$ is thus isomorphic to a general model of $\mathcal{A}$.

We will show in [BS 1] that any model of the elementary theory of countable ordinals, ET [co] , can be changed into a general model of MT[co] by specifying a field of subsets of its domain. Also obviously

(EXT) and (COMP) are true in all standard models, i.e. they belong to the theories we are interested in. Thus the restriction to general structures seems to be justified. We will therefore change our original definition, and from now on use "consequence" and all related notions including the operator S as if defined relative to general structures. Also we will fix the system $\mathcal{P}_2$ and say "derivable" short for "derivable in $\mathcal{P}_2$". We will write

$\mathfrak{A} \vdash \Upsilon$ for : Υ is derivable from $\mathfrak{A}$ (in $\mathcal{P}_2$),

$\mathfrak{A} \Vdash \Upsilon$ for : Υ is a (general) consequence of $\mathfrak{A}$.

Following the usage of Church [Ch 2], we define $\mathfrak{A}$ to be a <u>set of postulates</u> for the theory $\mathcal{T}$ iff exactly the sentences of $\mathcal{T}$ are derivable from $\mathfrak{A}$. By the completeness theorem, $\mathfrak{A}$ is a set of postulates for $\mathcal{T}$ iff $\mathfrak{A}$ is an axiom system for $\mathcal{T}$. Although we will work with derivations, we will prefer the latter term which is more familiar. One should keep in mind that, what in [Bü] is called an axiom system is a standard axiom system in our terminology.

The main result of this paper will be that, except for MT [α] where $\alpha < \omega^{\omega}$, the standard axiom systems of [Bü] are not axiom systems. In fact there are $\mathcal{L}$- expressible consequences of the axiom of choice which ascertain the existence of sets with certain properties. Thus these statements trivially hold in all standard structures, but they need not hold in general structures. See the end of section 2 , the beginning of section 7 and the section 8 for details.

The step from standard axiom systems to postulate systems is done in accordance with Skolem's Thesis which is discussed in the introduction. E.g. Dedekind's proof that Peano's axioms for the successor function on the natural numbers are categorical, amounts in our language to a proof that they are standard categorical: they admit (at most) one standard model. The hidden primitive in Peano's language is the ϵ - symbol; the set variables are intended to range over the full power set of the range of the individual variables, but this intention is not stated in the axioms. We know that we cannot force the full power set by axioms; but we have seen that if we add the axioms of extensionality and comprehension to an axiom system, we force the models to be general, i.e. the set variables range over actual sets of the model and ϵ becomes the membership relation.

One word of caution: Whereas for number theory, as for any complete theory, e.g. $MT[\alpha]$, any two models are elementarily equivalent, there are models of $MT[\omega]$ which are not elementarily equivalent to any of its standard models. Thus no axiomatization can do here any better.

A last remark concerning section 8 : We have defined $MT[\omega_1]_0$ as the set of sentences Σ s.t. $\Sigma \rightarrowtail [\omega_1]$. Since ω_1 is taken (as the first uncountable ordinal) from a certain model $\mathcal{M}$ of set theory, more precisely we should write $\omega_1^{\mathcal{M}}$, and thus $MT[\omega_1^{\mathcal{M}}]$. As this paper will show $MT[\omega_1^{\mathcal{M}}]$ depends on the model $\mathcal{M}$ we choose. Thus let $\mathcal{T}$ be any set theory. Define $MT[\omega_1^{\mathcal{T}}]$ as the intersection of all $MT[\omega_1^{\mathcal{M}}]$ where $\mathcal{M}$ is a model of $\mathcal{T}$. We will show that $MT[\omega_1^{ZFC}]$ is complete, i.e. for any two models $\mathcal{M}$ and $\mathcal{M}^*$ of ZFC , $MT[\omega_1^{\mathcal{M}}] = MT[\omega_1^{\mathcal{M}^*}]$. Therefore $MT[\omega_1^{ZFC}]$ is just the $MT[\omega_1]$ of [Bü] . The independence of the splicing axiom shows that $MT[\omega_1^{ZF}] \subsetneqq$ $\subsetneqq MT[\omega_1^{ZFC}]$; it is open whether $MT[\omega_1^{ZF}]$ is decidable. For details see section 8 .

For the careful reader an explanation might be in order. Formalizing metamathematics, normally one would define $MT[\omega_1]$ as the set of those sentences Σ s.t.

$$\mathcal{T} \vdash "[\omega_1] \rightarrowtail \Sigma" .$$

Here $\mathcal{T}$ is any suitable set theory, and $"[\omega_1] \rightarrowtail \Sigma"$ is the formalization of the statement $[\omega_1] \rightarrowtail \Sigma$ as a sentence of set theory. Since the structure $[\omega_1]$, however, is definable in set theory, it seems more natural to define $MT[\omega_1]$ as the set of those sentences Σ s.t.

$$\mathcal{T} \vdash \Sigma[\omega_1] .$$

Here $\Sigma[\omega_1]$ is the relativization of Σ to the set ω_1 , a notion introduced in the beginning of section 2 . It seems to be open whether the two definitions of $MT[\omega_1]$ coincide for ZF , say. Of course, this question can be asked for arbitrary structures $\mathfrak{D}$ instead of $[\omega_1]$, if one defines $\Sigma[\mathfrak{D}]$ in a suitable way. It seems that this problem has not yet been considered in model theory; this is strange since e.g. the former notion depends on the choice of an arithmetization whereas the latter does not.

2. Axiom systems for MT [co] and for MT [α] , $\alpha < \omega_1$

In this section we will extend the standard axiom systems of the preceding paper [Bü] to get axiom systems for the theories MT [co] and MT [α] for $\alpha < \omega_1$. Since [Bü] will be quoted extensively, we will simply write e.g. "section [Bü] 1", "lemma [Bü] 1.1". We will take over most of the terminology used in [Bü] , cf. especially the first and the last paragraphs of section [Bü] 1 . So WLO will denote the axiom of well-order, i.e. the universal closure of the cojunction of the following formulas:

$$\neg \; x < x,$$

$$x < y \wedge y < z \longrightarrow x < z,$$

$$x < y \vee y < x \quad \vee \quad x = y,$$

$$(\exists z)Xz \longrightarrow (\exists z)\left[Xz \wedge (\forall y)^{z} \neg Xy\right].$$

Obviously, any standard model of WLO is well-ordered, and thus can be identified with an ordinal.

Let Σ be a formula not containing the variables x and X. The notion of the relativization $\Sigma[x]$, $\Sigma\{x\}$ and $\Sigma[X]$ of Σ to the upper bound x , to the lower bound x and to the subset X resp. will play an important role in this paper; see the end of section [Bü] 1. We get $\Sigma[X]$ by replacing all quantifiers of the form

$$(\overset{\forall}{\exists} z) \text{ by } (\overset{\forall}{\exists} z \in X) \text{ and } (\overset{\forall}{\exists} Z) \text{ by } (\overset{\forall}{\exists} Z \subseteq X).$$

Then

$$\Sigma[x] \equiv_{df} \Sigma[\{t;\, t < x\}] ,$$

$$\Sigma\{x\} \equiv_{df} \Sigma[\{t;\, x \leq t\}] .$$

(If we substitute an expression for a free variable, this is understood as an abbreviation, which can be replaced by a correct formula.) Note that free variables are not relativized.

Examples:

$$(\exists t)\, Ut\,[x]\,[X] \equiv (\exists t \in X)_0^x\, Ut\ ,$$

$$((\forall y)\,(\exists z)\, y' = z)\,[U] \equiv (\forall y \in U)\,(\exists z \in U)\,(y < z\ \wedge$$

$$\neg(\exists t \in U)\,[y < t \wedge t < z])\ ,$$

$$\phi\,[x]\,\{y\} \longleftrightarrow \phi\,\{y\}\,[x]\ .$$

Further if $\alpha < \beta < \gamma$, and α, β are definable, then

$$\Sigma\,\{\alpha\}\,[\beta] \succ\!\!- [\gamma]$$

iff Σ is true in the half-open interval $[\alpha, \beta)$.

Let

$$\mathfrak{D} = \langle D, \mathcal{P}_0(D)\ ;\ \in\ ,\ =\ ,\ 0 \rangle$$

be any general model, let $U \in \mathcal{P}_0(D)$. Then the <u>model</u> $\mathfrak{D}$ <u>relativized to</u> U is

$$\mathfrak{D}\,[U] =_{df} \langle U\ ,\ \{Y \wedge U\ ;\ Y \in \mathcal{P}_0(D)\}\ ;\ \in\restriction U\ ,\ =\restriction U\ ,\ 0\restriction U \rangle\ .$$

Obviously, $\mathfrak{D}\,[U]$ is a general model, and for any sentence Σ (if we allow a somewhat sloppy notation)

$$\Sigma \succ\!\!- \mathfrak{D}\,[U] \quad \text{iff} \quad \Sigma\,[U] \succ\!\!- \mathfrak{D}\ .$$

As an exercise in relativization we write down the definition of limit numbers of higher order, given in Büchi [Bü 3] , p.4. There all numbers are limit numbers of order 0 ; limit numbers of order p + 1 are limits of limit numbers of order p .

$$LM \equiv_{df} (\forall y)\,(\exists z)\ y < z\ ,$$

$$Lm(x) \equiv_{df} LM\,[x] \wedge x \neq 0\ ,$$

$$LM_0 \equiv_{df} (\forall z)\ z = z\ ,$$

$$Lm_p(x) \equiv_{df} LM_p\,[x]\ ,$$

$$LM_{p+1} \equiv_{df} LM\,[Lm_p]\ ,$$

$$lm_p \equiv_{df} LM_p \wedge \neg\, LM_{p+1}\ ,$$

$$lm_p(x) \equiv_{df} lm_p[x] .$$

Thus $Lm(x)$, $Lm_p(x)$, $lm_p(x)$ say that x is a limit number, is zero or a limit number of order at least p , and is a limit number of order exact p , resp. The absolute forms say the same about the order type of the domain in which they are true. One shows easily the equivalence of this definition of $Lm_{p+1}(x)$, i.e.

$$(\forall y \in Lm_p)^x (\exists z \in Lm_p)^x \; y < z ,$$

with the one given in Büchi [Bü 3] , namely

$$(\forall y)^x (\exists z \in Lm_p)^x \; y < z .$$

Note that lm_o is equivalent to Suc, end of section [Bü]1. Zer is always false, since by definition a structure has a non-empty domain. Thus the ordinal 0 is not covered by our treatment. Of course, every model of WLO contains a first element 0 , defined by Zer[x]. The successor relation Scs , however, need not define a function, since the model may have a last element. We will nevertheless normally use the function symbol ' , and will give hints at points where one has to be careful. We can easily derive from WLO :

$$(\forall y)^x (\exists z) \; Scs(y,z) .$$

We also note some properties of limit numbers:

$$LM_1 \longleftrightarrow LM ,$$

$$Lm_1(x) \longleftrightarrow LM[x] ,$$

$$Lm_1(x) \not\longleftrightarrow Lm(x) ,$$

$$LM_{p+1} \longrightarrow LM_p ,$$

$$Lm_{p+1}(x) \longrightarrow Lm_p(x) .$$

For any $n < \omega$, we define y <u>is congruent</u> z <u>modulo</u> n as

$$y \equiv z(n) \equiv_{df} (\forall U) \left[Uy \wedge (\forall t) \left[Ut \longleftrightarrow U(t+n) \right] \longrightarrow Uz \right] .$$

This is the same definition as for the integers. Note that $y \equiv z(n)$ for $n \neq 1$ implies that there is no limit number between y and z, i.e.

$$y \leq z < y+\omega \quad \text{or} \quad z \leq y < y+\omega .$$

The reader is advised to write down the formula

$$Lm_j(y) \wedge Lm_j(z) \wedge (y \equiv z(n)) \; [Lm_j]$$

in more detail.

We define

$Cof(U) \equiv_{df} (\forall y)\,(\exists z \in U)\; y \leq z$, U <u>is</u> <u>cofinal.</u>

Note that in a domain of successor type, U is cofinal iff it contains the last element. According to the definition of cof(U) at the end of section [Bü]1 , cofinal sets are always of limit type; thus only domains of limit type have cofinal subsets. We have changed the definition slightly, in order to facilitate a parallel discussion of the successor case and the limit case. E.g. let

$$ACC \equiv_{df} (\exists U).\; Cof\,(U) \wedge \neg LM_2[U] .$$

Then, using WLO , ACC is equivalent to $Acc_0 \vee Suc$ of [Bü], end of section 1 , and also to

$$(\exists U) \;.\; Cof(U) \wedge \neg(\exists x)Lm(x) \; [U] .$$

Together with the remarks at the end of section [Bü]1 we get:

<u>Theorem 2.1*:</u> The set

$$\mathcal{C}_0 =_{df} \{WLO \;,\; ACC \;,\; (\forall x)ACC[x]\}$$

is a finite standard axiom system for MT[co].

We note some trivial facts about $\mathcal{C}_0$. These statements will become non-trivial if we replace "standard consequence" by "consequence" (see theorem 2.2).

<u>Theorem 2.2*:</u> $(\forall X \neq \emptyset)\; \mathcal{C}_0[X]$ is a standard consequence of $\mathcal{C}_0$.

<u>Proof:</u> Trivial by theorem 2.1*, since every subset of a countable well-ordered set is countable and well-ordered. Q.e.d.

<u>Corollary 2.1*:</u> ϕ is a standard consequence of $\mathcal{C}_0$ iff $(\forall X{=}\emptyset)\ \phi[X]$ is.

<u>Proof:</u> The "if"-part is trivial with $X=\{t;\ t=t\}$. For the "only if"-part let $\mathcal{D}$ be a standard model of $\mathcal{C}_0$, let $\emptyset \neq X \in \mathcal{P}_0(\mathcal{D})$. By theorem 2.2* , the relativized model $\mathcal{D}[X]$ is a standard model of $\mathcal{C}_0$. Therefore $\phi \succ\!\!- \mathcal{D}[X]$, and thus $\phi[X] \succ\!\!- \mathcal{D}$, which proves

$$(\forall X \neq \emptyset)\ \phi[X] \succ\!\!- \mathcal{D}$$ Q.e.d.

<u>Corollary 2.2*:</u> ϕ is a standard consequence of $\mathcal{C}_0$ iff $(\forall x{>}o)\phi[x]$ is iff $(\forall x)\ \phi\{x\}$ is.

<u>Proof:</u> The "only if"-part follows from corollary 2.1*. For the "if"-part note that for any countable ordinal α, $\alpha+1$ is again a countable ordinal. The implication concerning $(\forall x)\ \phi\{x\}$ is trivial. Q.e.d.

In the unstarred analogues to these statements, i.e. theorem 2.2 and corollaries 2.1+2, the word "standard" is cancelled. (Theorem 2.1 will follow later.)

<u>Theorem 2.2:</u> $\mathcal{C}_0 \vdash (\forall X \neq \emptyset)\ \mathcal{C}_0[X]$.

<u>Proof:</u> WLO[X] is trivial. We have to derive the relativized accessibility statements. So let $X \neq \emptyset$ be given. To show ACC[X] , we have to find a set U such that:

$$\begin{aligned} (1)\qquad & U \subseteq X \wedge (\forall y{\in}X)\ (\exists z{\in}U)\ y < z\ \wedge \\ & \wedge\ \neg(\exists x{\in}U)\ \left[(\exists y)^X Uy \wedge (\forall y{\in}U)^X(\exists z{\in}U)^X y < z\right] \end{aligned}$$

(see the remark in front of theorem 2.1*). We will distinguish two cases:

Case (a): X is cofinal. By ACC there is a cofinal ω-sequence V , i.e.

$$(2)\qquad \mathrm{cof}(V) \wedge \neg(\exists x)\mathrm{Lm}(x)\ [V]\ .$$

Define U by

(3) $$Uz \longleftrightarrow Xz \wedge (\exists v \in V)\ v \leq z \wedge (\forall t)_v^z \neg Xt \ .$$

We will show that U is a cofinal ω-sequence in X .
$U \subseteq X$ is trivial. Now let $y \in X$ be given. Since V is cofinal, there is $v \in V$, $y \leq v$. Again since X is cofinal, there is $z \in X$, $v \leq z$. If we choose z minimal, we have $z \in U$. This proves Cof (U) [X] , i.e. the second part of (1). To prove the last part of (1), note that by (3)

(4) $$Uy \wedge Uz \wedge y<z \longrightarrow (\exists v \in V)\ y<v \leq z \ .$$

Now let y_0 be first element of U , let $x \in U$ where $y_0 < x$. Assume x to be a limit of U , i.e.

(5) $$(\forall y \in U)^x (\exists z \in U)^x\ y<z \ .$$

We want to prove that then V contains a limit. Let w be minimal s.t.

(6) $$w \leq x \wedge (\forall t)_w^x \neg Ut \ .$$

By minimality of w ,

(7) $$(\forall y)^w (\exists z \in U)^w y \leq z \ .$$

Now let $u \in V$, $u < w$ be given. By (7) there is $y \in U$, $u \leq y < w$. By (5) and (6) there is $z \in U$, $y < z < w$. By (4) there is $v \in V$, $y < v \leq z$. Thus we have proved

(8) $$(\forall u \in V)^w (\exists v \in V)^w\ u < v \ .$$

Let $v \in V$ be minimal s.t. $w \leq v$ (V is cofinal). Since by (3) and (6) v is not the first element of v , we get Lm(v)[V] , which contradicts (2). Therefore $\neg(\exists x)$Lm(x) [U] , what proves (1).
Case (b): X is not cofinal. Then there is a minimal w s.t.

$$(\forall z \in X)\ z < w \ .$$

Since w is minimal, we have Cof(X)[w] . If we relativize the proof of case (a) to w , we get again ACC[X] , using this time ACC[w] instead of ACC .

Finally, $(\forall x \in X) ACC[x][X]$ is proved similarly, which proves the theorem. Q.e.d.

<u>Corollary 2.1:</u> $\mathcal{C}_0 \vdash \phi$ iff $\mathcal{C}_0 \vdash (\forall X \neq \emptyset)\, \phi[X]$

<u>Corollary 2.2:</u> $\mathcal{C}_0 \vdash \phi$ iff $\mathcal{C}_0 \vdash (\forall x > o)\ \phi[x]$
iff $\mathcal{C}_0 \vdash (\forall x)\ \phi\{x\}$

<u>Proof:</u> We get the corollaries and their proofs by removing stars and the word "standard" from the corollaries 2.1* and 2.2*. Only for the proof of corollary 2.2 we have to generalize the argument involving $\alpha+1$: Let

$$\mathcal{D} = \langle D\ ,\ \mathcal{P}_o(D)\ ;\ \in\ ,\ =\ ,\ <\ \rangle$$

be any general model of $\mathcal{C}_0$. Enlarge $\mathcal{D}$ by an element, added at the end. I.e. let $a \notin D$, let

$$\mathcal{D}^a =_{df} \langle D \cup \{a\}\ ,\ \mathcal{P}_o(D) \cup \{Y \cup \{a\}\ ;\ Y \in \mathcal{P}_o(D)\}\ ;\ \in_{\mathcal{D}^a}\ ,$$
$$=_{\mathcal{D}^a}\ ,\ <_{\mathcal{D}} \cup \{(y,a)\ ;\ y \in D\}\ \rangle .$$

Then obviously $WLO \succ\!\!- \mathcal{D}^a$, since $WLO \succ\!\!- \mathcal{D}$.

Further

$$Cof(\{a\}) \wedge \neg IM[\{a\}] \succ\!\!- \mathcal{D}^a ,$$

thus $ACC \succ\!\!- \mathcal{D}^a$.

Since

$$(\forall x)\ ACC\ [x] \wedge ACC \succ\!\!- \mathcal{D} ,$$
$$(\forall x)^a ACC\ [x] \wedge ACC[a] \succ\!\!- \mathcal{D}^a ,$$
$$\text{i.e.} (\forall x)\ ACC\ [x] \succ\!\!- \mathcal{D}^a , \text{ and thus } \mathcal{C}_0 \succ\!\!- \mathcal{D}^a .$$

Therefore $\mathcal{C}_0 \vdash (\forall x > o)\, \phi[x]$ implies $(\forall x > o)\, \phi[x] \succ\!\!- \mathcal{D}^a$, especially $\phi[a] \succ\!\!- \mathcal{D}^a$, and thus $\phi \succ\!\!- \mathcal{D}$. Since $\mathcal{D}$ was an arbitrary model of $\mathcal{C}_0$, we have $\mathcal{C}_0 \vdash \phi$. Q.e.d.

The fact should be stressed that we have used the completeness theorem to prove corollaries 2.1 + 2.2. This seems to be essential, since the stronger statements

(1) $\quad \phi \longleftrightarrow (\forall X \neq \emptyset)\ \phi[X]\ ,\ \phi \longleftrightarrow (\forall x > o)\ \phi[x]$

are wrong, even for sentences ϕ . (The example of the model $[\omega]$ and of the sentences LM resp. ¬LM shows that (1) is not even standard-valid.) Therefore one cannot use induction on the structure of ϕ to prove the corollaries.

To get standard axiom systems for $MT[\alpha]$, $\alpha < \omega_1$, we introduce abbreviations which are used in unpublished work of Mostowski and Tarski on the elementary theory of ordinals, see Mostowski - Tarski [MT 2].

$S_k \equiv_{df} (\exists x)\ lm_k(x)$, $k \geqslant o$, there are limit numbers of order k .
$T_{k,q} \equiv_{df} (\exists x_1,\dots,x_q \in Lm_k)\left[x_1 < \dots < x_q \wedge (\forall y \in Lm_{k+1})\ y < x_1\right]$,
there are at least q <u>final limit numbers of order</u> k .

By this definition, final limit numbers of order k occur after the last Lm_{k+1} . S_k and $T_{k,q}$ are defined for any $k \geqslant o$, $q \geqslant o$; note that $T_{k,o}$ is always true. To get acquainted with these notions the reader may prove:

<u>Lemma 2.1*:</u> For any $k,q \geqslant o$ the following is true:

a) $\quad S_{k+1} \longrightarrow S_k$

b) $\quad T_{k,q} \longrightarrow S_k$, if $q > o$

c) $\quad T_{k,q+1} \longrightarrow T_{k,q}$

d) $\quad S_k \wedge \neg S_{k+1} \longrightarrow T_{k,1}$

For any $o \leqslant l \leqslant k < \omega$ and any sequence $n_k,\dots,n_l$, $o \leqslant n_i < \omega$ for $i = l,\dots,k$, we define

$$\Sigma^{l}_{n_k,\dots,n_l} \equiv_{df} \bigwedge_{i=l}^{k} \left[T_{i,n_i} \wedge \neg T_{i,n_i+1}\right].$$

Thus e.g. $\Sigma^2_1 \equiv T_{2,1} \wedge \neg T_{2,2}$, whereas $\Sigma^1_{1,2} \equiv T_{1,2} \wedge \neg T_{1,3} \wedge T_{2,1} \wedge \neg T_{2,2}$.

Let $o \leq l \leq k$, let $o \leq n_i$ for $i=1,\dots,k$, let $o < q$, let α be a countable ordinal of $k+1$-character

$$\langle (\mu),\ m_k,\dots,m_{l+1},\ m_l+1,\ o,\dots,o \rangle ,$$

i.e.

$$\alpha = \omega^{k+1}\cdot\mu + \sum_{i=k}^{l} \omega^{i}\cdot m_i + \omega^{l}.$$

(For the definition of <u>character</u>, <u>expansion</u>, <u>head</u> and <u>tail</u> see section [Bü]4, after theorem 4.8.)

$$S_k \succ\!\!-\ [\alpha] \text{ iff } \alpha \geq \omega^k$$
$$\text{iff } \mu \neq o \text{ or } m_k \neq o ,$$

$$T_{k,q} \succ\!\!-[\alpha] \text{ iff either } m_i = o \text{ for } i = 1,\dots, k-1 \text{ and } q < m_k$$
$$\text{or } m_i \neq o \text{ for some } 1 \leq i < k \text{ and } q \leq m_k,$$

$$\textstyle\sum^{l}_{n_k,\dots,n_l} \wedge LM_l \succ\!\!-\ [\alpha] \text{ iff } m_i = n_i \text{ for } i = 1,\dots,k .$$

Finally, for $\alpha = \sum_{i=k}^{l} \omega^{i}\cdot n_i + \omega^{l}$ we write

$$\textstyle\sum_\alpha \equiv_{df} \sum^{l}_{n_k,\dots,n_l} \wedge LM_l \wedge \neg S_{k+1} .$$

Then obviously for any β

$$\textstyle\sum_\alpha \succ\!\!-\ [\beta] \text{ iff } \beta = \alpha .$$

Thus we get standard axiom systems for $MT[\alpha]$ as follows: Let

$$\alpha = \omega^{\omega}\cdot\nu + \sum_{i=k}^{l} \omega^{i}\cdot n_i + \omega^{l}\cdot n,\ 0 \leq l \leq k < \omega,\ 0 \leq n_i < \omega,\ n = 0,1 .$$

Assume k to be choosen as small as possible, i.e. $n_k \neq 0$ if $k > l$, and $n = 1$ if $n_i \neq o$ for some i .

<u>1.case:</u> $\nu = 0$, $n = 1$; i.e. $0 < \alpha < \omega^\omega$. Let

$$A_\alpha =_{df} \mathcal{C}_o \cup \{ \textstyle\sum_\alpha \}$$

<u>2.case:</u> $\nu \neq 0$, $n = 1$; i.e. $\omega^\omega < \alpha$, $\alpha \neq \omega^\omega\cdot\mu$. Let

$$A_\alpha =_{df} \mathcal{C}_o \cup \{ \textstyle\sum^{l}_{n_k,\dots,n_l} \wedge LM_l \} \cup \bigcup_{j>k} \{ S_j \wedge \neg T_{j,1} \} .$$

<u>3.case:</u> $\nu \neq 0$, $n = 0$; i.e. $\alpha = \omega^\omega\mu$ for some $\mu \neq 0$. Let

$$A_\alpha =_{df} \mathcal{C}_o \cup \bigcup_{j<\omega} \{ LM_j \} .$$

Theorem 2.3*: For any ordinal α, $0<\alpha<\omega_1$, $\mathcal{A}_\alpha$ is a complete standard axiom system for MT[α]. For $\alpha<\omega^\omega$, $\mathcal{A}_\alpha$ is finite. For $\alpha\geq\omega^\omega$, MT[α] is not finitely axiomatizable.

Note that for $\alpha<\omega^\omega$ $\mathcal{A}_\alpha$ is standard categorical: [α] is its only standard model. For $\alpha\geq\omega^\omega$ the situation is different: We know from theorem [Bü]4.9 that for $\alpha,\beta\geq\omega^\omega$, [$\alpha$] and [$\beta$] are elementarily equivalent iff they have the same ω-tail. (We can write "elementarily equivalent" instead of "MT-equivalent", since we have altered [α] into a standard structure which requires MT-formulas for interpretation.) The last part of the theorem follows from a remark following theorem [Bü] 4.8':

Remark 2.1 : If a sentence has a countable model, then it has a model [α] where $\alpha<\omega^\omega$. I.e. $\mathrm{MT}[\mathrm{co}] = \mathrm{T}\big[\{[\alpha];\ \alpha<\omega^\omega\}\big]$.

We note a consequence which will be very useful later:

Corollary 2.3: α is definable in MT[co] iff $\alpha<\omega^\omega$. A defining formula for $\alpha<\omega^\omega$ is $\sum_\alpha[v]$. (Here we add the definition $\sum_0 \equiv_{df} \neg(\exists t)\,t=t$.)

It would be nice if the standard axiom systems we have so far established were also axiom systems. This is, however, not the case. There are weak forms of the axiom of choice which hold in all standard models of $\mathcal{C}_0$, but fail in other models of $\mathcal{C}_0$.

The axiom of choice we need reads as follows: To any collection of non-empty, disjoint sets of sets there is a choice set which contains exactly one set from each member of the collection. This is of course not expressible in the language of MT[co]. Therefore we think the collection be given by a formula $\phi(y,z,Z)$ and a set U s.t. for any neighbouring elements $y<z$ of U the set

$S_{y,z} = \{ Z \subseteq [y,z) \; ; \; \phi(y,z,Z) \}$ is a member of the collection. Then the axiom states that there is a set Z s.t. for any two neighbours $y < z$ of U, $Z \cap [y,z) \in S_{y,z}$. I.e. the small sets $Z \in S_{y,z}$ are spliced together to give the choice set. It is easier if $Z \in S_{y,z}$ is not restricted to $[y,z)$, but rather $\phi(y,z,Z)$ does not depend on the values of Z outside $[y,z)$.

We define

$$Z \overset{z}{\underset{y}{=}} Y \equiv_{df} (\forall t)_y^z \, Zt \equiv Yt \, .$$

Then SPLICE is the following axiom schema:

For any formula $\phi(y,z,Z)$ satisfying

(1) $$y < z \wedge Y \overset{z}{\underset{y}{=}} Z \rightarrow [\phi(y,z,Y) \longleftrightarrow \phi(y,z,Z)]$$

the following shall be an axiom:

(2) $$(\forall y,z \in U)[(z = y')\,[U] \longrightarrow (\exists Z)\,\phi(y,z,Z)] \longrightarrow$$
$$\longrightarrow (\exists Z)\,(\forall y,z \in U)[(z = y')\,[U] \longrightarrow \phi(y,z,Z)] \, .$$

Note that (1) is rather strong, since it does not depend on U. This form of SPLICE, however, is enough for our purposes. Note also that it would be enough to restrict SPLICE to formulas $\phi(y,z,Z^k)$ where $k = 1$. We leave it as an exercise to the reader to show by induction on k that the general axiom is derivable from the restricted form.

Let

$$\overline{\mathcal{C}_o} =_{df} \mathcal{C}_o \cup \{\text{SPLICE}\} \, ,$$
$$\overline{\mathbb{A}}_\alpha =_{df} \mathbb{A}_\alpha \cup \{\text{SPLICE}\} \, .$$

<u>Theorem 2.1:</u> If we assume the axiom of choice, then $\overline{\mathcal{C}_o}$ is an axiom system for $MT[co]$.

<u>Theorem 2.3:</u> For any ordinal α, $0 < \alpha < \omega_1$, $\overline{\mathbb{A}}_\alpha$ is a complete axiom system for $MT[\alpha]$. For $\alpha \geq \omega^\omega$ this involves the axiom of choice.

The proof of these theorems will extend through the next four sections. In section 7 we will present an axiom system for ω_1.

The independence of SPLICE from $\mathcal{C}_0$ will be shown in section 8 . SPLICE is derivable, however, from $\mathcal{A}_\alpha$ for $\alpha < \omega^\omega$; see section 4 . The role of AC in theorems 2.1+3 will also be discussed in section 8 . The reader may prove as an exercise that theorem 2.1 , and thus corollaries 2.1+2 extend from $\mathcal{C}_0$ to $\overline{\mathcal{C}}_0$.

The proof in section 5 will show that we need SPLICE only for kernels of automata formulas, i.e. for formulas $\phi(y,z,Z)$ of the form

$$(3)\quad Zy = d \wedge (\forall t)_y^z \, H[Xt,\ Zt,\ Zt'] \wedge (\forall x)_y^{z'} \, \mathcal{K}[\sup^x Z,\ Zx] \\ \wedge \{Zt;\ y \leq t < z\} = D \wedge Zz = d \ .$$

Such formulas satisfy

$$(1')\quad y < z \wedge Y \overset{z'}{\underset{y}{=}} Z \longrightarrow [\phi(y,z,Y) \longleftrightarrow \phi(y,z,Z)] \ ;$$

thus we could weaken SPLICE by requiring (2) for formulas (3) only. Theorem 2.1 shows, however, that this weak form is as strong as our SPLICE.

3. The automata normal form

We will prove the theorems of section 2 on axiom systems for MT[α] and MT[co] in the following way: We will go through the decidability proofs of [Bü] for these theories, and will show that all the equivalences which justify the steps of the decision procedure are derivable. Therefore every sentence of MT[α] is $\bar{A}_\alpha$-derivably equivalent to "true" or "false", which proves theorem 2.3. Using theorem 2.3 we will prove theorem 2.1.

First we need a chain of lemmata on derivations. In this section, derivations will use only the axioms of well-order, WLO. The following are trivial:

Lemma 3.1: The formulas of lemma 2.1* are derivable from WLO.

Lemma 3.2: The normal form of Lemma [Bü]1.1 is derivable (within $\mathfrak{P}_2$), i.e. to any formula Σ one can find a formula Φ in normal form such that $\vdash \Sigma \longleftrightarrow \Phi$

We call a set transitive if it contains with any ordinal all smaller ordinals:

$$\mathrm{Trans}(X) \equiv_{df} (\forall z)\left[Xz \longrightarrow (\forall y)_0^z\, Xy\right].$$

We take from [Bü] the convention to reserve the variable x for limit numbers. We note several forms of the induction principle, which are known to be equivalent with the minimum principle:

Lemma 3.3: From WLO are derivable:

a) $(\forall z)\left[(\forall y)^z Xy \longrightarrow Xz\right] \longrightarrow (\forall z)Xz$,

b) $Xo \wedge (\forall t)\left[Xt \longrightarrow Xt'\right] \wedge (\forall x)\left[(\forall y)^x Xy \longrightarrow Xx\right] \longrightarrow (\forall z)Xz$,

c) $Xo \wedge \mathrm{Trans}(\neg X) \longrightarrow (\forall z)Xz$.

Lemma 3.4: Pigeon hole principle:

$$\mathrm{WLO} \vdash (\forall y \in Y) \bigvee_{i=1}^{n} \Phi_i(y) \longrightarrow (\exists Z).\ \mathrm{Cof}\ (Z)[Y] \wedge \bigvee_{i=1}^{n} (\forall z \in Z)\, \Phi_i(z)$$

<u>Proof:</u> Define sets Z_i , $i = 1,\dots,n$ by

$$Z_i t \longleftrightarrow Yt \wedge \phi_i(t) .$$

Assume: $\neg \mathrm{Cof}(Z_i)[Y]$, $i = 1,\dots,n$. Then

$$(\exists z_1,\dots,z_n \in Y) \bigwedge_{i=1}^{n} (\forall y)_{z_i} \neg Z_i y ,$$

and therefore by ordering the z_i

$$(\exists z \in Y) \bigwedge_{i=1}^{n} (\forall y)_z \neg Z_i y ,$$

which is a contradiction to $Y = \bigcup_{i=1}^{n} Z_i$. Thus we have derived

$$\bigwedge_{i=1}^{n} (\forall t)\left[Z_i t \longleftrightarrow Yt \wedge \phi_i(t)\right] \wedge (\forall y \in Y) \bigvee_{i=1}^{n} \phi_i(y) \rightarrow \bigvee_{i=1}^{n} \mathrm{Cof}(Z_i)[Y].$$

If we apply existential quantifiers to both sides of the implication, and use (COMP), we get the lemma. Q.e.d.

Note that the lemma becomes trivial (but the proof remains unchanged), if Y has a last element. We have given a rather detailed account of the proof, since similar considerations, which are typical in finite automata theory, will occur over and over again. It should be easy to set up a completely formal derivation schema for the lemma from the given proof. The reader might consult section I.5.a of Siefkes [Si1] for more details.

<u>Lemma 3.5:</u> From WLO are derivable

a) $\mathrm{Trans}\,(X) \longleftrightarrow (\forall t)\left[Xt' \longrightarrow Xt\right] \wedge (\forall x)\left\{Xx\right\} = \sup^x X$.

b) $y < z \longleftrightarrow (\exists X \in \mathrm{Trans})\left[Xy \wedge \neg Xz\right]$.

Here we write $\mathfrak{U}\left[\sup^x Z\right]$, where $\mathfrak{U}$ is a propositional formula, short for

$$\bigvee_{\{S\ ;\mathfrak{U}[S]\}} \left[\bigwedge_{s \in S} (\exists^x t)\ Zt = s \wedge \bigwedge_{s \notin S} \neg(\exists^x t)\ Zt = s\right].$$

<u>Lemma 3.6:</u> Lemma Bü 1.2 is derivable from WLO.

<u>Proof:</u> Part c) and the fact that Boolean combinations of sup-condi-

tions are again sup-conditions, are immediate from our notation. $(\exists^x t)L[Zt]$ is by the pigeon hole principle equivalent to

$$\bigvee_{\{s;L[s]\}} s \in \sup{}^x Z ,$$

which yields part a). If $Z = (Z_1,\ldots,Z_m)$ and $\tilde{Z} = (Z_1,\ldots,Z_n)$ where $m < n$, then any condition $\mathfrak{U}[\sup^x Z]$ can be restated in the form $\mathfrak{U}[\sup^x \tilde{Z}]$: just replace Z by $\tilde{Z}$ in the definition. This yields part b). Q.e.d.

Lemma 3.7: The equivalences of lemmata [Bü]1.3+4 are derivable from WLO.

Definition of sets by (primitive) recursion as in lemma [Bü]1.3 plays an important rôle throughout the decision procedure. Therefore it is essential to derive that recursively defined sets (are unique and) exist. We will write

$$(\exists !! Z)\ \Psi(Z)$$

short for

$$(\exists Z)\ [\Psi(Z) \wedge (\forall Y)\ (\Psi(Y) \longrightarrow Y = Z)] .$$

Lemma 3.8: Primite recursion. Let $\phi(y,Zt;t<y)$ be a formula in which for all prime formulas of the form $Z_i t$ the variable t is bound by t. Then

$$\mathrm{WLO} \vdash (\exists !! Z)\ (\forall y)\ [Zy \longleftrightarrow \phi(y,Z)] .$$

Proof: Recall from the end of section 2 that

$$X \overset{z}{\underset{y}{=}} Z \text{ is short for } (\forall v)_y^z\ [Xv \longleftrightarrow Zv] .$$

Note that

(1) $$X \overset{z}{\underset{0}{=}} Z \longrightarrow [\phi(z,X) \longleftrightarrow \phi(z,Z)]$$

is derivable within $\mathcal{P}_1$ by induction on the structure of ϕ. We define a formula Ψ by

(2) $$\Psi(z,X) \equiv_{df} (\forall y)_0^z\ [Xy \longleftrightarrow \phi(y,X)] ,$$

and show by induction on z :

(3) $\quad \Psi(z,X) \wedge \Psi(z,Z) \longrightarrow X \overset{z}{\underset{0}{=}} Y$.

Let (3) be proved for all $z < u$. To prove it for u , suppose

(4) $\quad \Psi(u,X) \wedge \Psi(u,Z)$.

From (4) and the induction hypothesis we get

(5) $\quad (\forall z)_0^u \; X \overset{z}{\underset{0}{=}} Z$.

If u is zero or a limit number, this yields directly $X \overset{u}{\underset{0}{=}} Z$. So let $u = v'$ for some v . From (5) we get $X \overset{v}{\underset{0}{=}} Z$, and thus by (1)

$$\phi(v,X) \longleftrightarrow \phi(v,Z) .$$

Therefore by (4) $Xv \longleftrightarrow Zv$, which again yields $X \overset{u}{\underset{0}{=}} Z$ from (5). This proves (3). To show the lemma we define Z by

(6) $\quad Zy \longleftrightarrow (\exists X) \left[\Psi(y',X) \wedge Xy\right]$.

We want to show $\Psi(u',Z)$ by induction on u . So let $\Psi(v',Z)$ be proved for all $v < u$. First suppose Zu . Then there is an X s.t.

(7) $\quad \Psi(u',X) \wedge Xu$.

From (3) and the induction hypothesis follows easily $X \overset{u}{\underset{0}{=}} Z$, and thus by (1)

$$\phi(u,X) \longleftrightarrow \phi(u,Z) .$$

Together with (7) this yields $\phi(u,Z)$. Thus we have proved

(8) $\quad Zu \longrightarrow \phi(u,Z)$.

To get the converse define X by

$$Xy \longleftrightarrow y = u \vee \left[y < u \wedge Zy\right] .$$

Then for all $v \leq u$

(9) $\quad X \overset{v}{\underset{0}{=}} Z$,

and thus by (1) for all $v \leq u$

(10) $\quad \phi(v,X) \longleftrightarrow \phi(v,Z)$.

Further by induction hypothesis for all $v < u$

(11) $\quad Zv \longleftrightarrow \Phi(v,Z).$

(9),(10) and (11) yield $\Psi(u,X)$. Now suppose $\Phi(u,Z)$. (10) implies $\Phi(u,X)$, which yields $\Psi(u',X)$ from Xu and $\Psi(u,X)$. From $\Psi(u',X)$ and Xu follows Zu by (6). This proves

$$\Phi(u,Z) \longrightarrow Zu$$

and thus by (8) and the induction hypothesis $\Psi(u',Z)$. (Note that the argument works for $u = o$ as well, since $\Phi(o,Z)$ does not depend on Z.) We have proved

$$(\forall u)\ \Psi(u',Z),$$

which implies

(12) $\quad (\forall y)\left[Zy \longleftrightarrow \Phi(y,Z)\right].$

The existence of Z follows from COMP , the uniqueness of a Z satisfying (12) follows from (3). Q.e.d.

<u>Remark:</u> We have stated and proved the recursion theorem for a single set variable. The extension to simultaneous recursion, i.e. to a string of set variables, is trivial.

The formulas which express in [Bü] the acceptance condition for finite automata work for successor ordinals only. We have to cover the limit case, too. Therefore, for a tupel Z of set variables we define the <u>global supremum</u> $\sup Z$ as the unrelativized version of the local supremum $\sup^x Z$:

$$\sup Z =_{df} \{a\ ;\ (\forall t)\ (\exists y)_t\ Zy = a\} = \{a\ ;\ \mathrm{Cof}\ (\{y\ ;\ Zy = a\})\} .$$

By corollary 2.2 , lemma 3.6 extends to the global supremum. Note that in a model $[\alpha']$, $\sup Z$ is just $\{Z\alpha\}$; in a model $[\alpha]$ of limit type, $\sup Z$ is the set of those states which occur cofinal in α .

This leads to the

Definition: a) A formula $\Psi(X)$ is in automata normal form iff Ψ is of the form

$$(\exists Z).\ E[Z_0] \wedge (\forall t)\ H[Xt,\ Zt,\ Zt'] \wedge (\forall x)\ \mathcal{K}[\sup^x Z,\ Zx] \wedge$$
$$\wedge\ \mathcal{L}[\sup X\ ,\ \sup Z]\ ,$$

where $\mathcal{L} = (L_0, \mathcal{L}_1)$ such that

$$\mathcal{L}[\sup X,\ \sup Z] \equiv_{df} (\exists y) \left\{(\forall t)\ t \leq y\ \wedge L_0\ [Xy,\ Zy]\right\} \vee$$
$$\vee \left\{LM \wedge \mathcal{L}_1\ [\sup Z]\right\}.$$

b) The automata normal form is deterministic if H and K are of the resp. forms:

$$(\forall t)\ Zt' = F\,[Xt,\ Zt]\ ,$$
$$(\forall x)\ Zx = \mathcal{G}\,[\sup^x Z]\ .$$

Note that, to be more precise, H has to be written as

$$(\forall t)\ (\forall y)\ [Scs\,(t,y) \longrightarrow H\ [Xt,\ Zt,\ Zy]]\ .$$

Note also that the final condition $\mathcal{L}_1$ for the limit case does not depend an X . The following lemma gives an example of an automata formula written unabbreviated:

Lemma 3.9: $WLO \vdash (\exists z)\ Xz \longleftrightarrow$:

$$\longleftrightarrow : (\exists Z).\ Zo\ \wedge (\forall t)\ (\forall y)\ [Scs\,(t,y) \longrightarrow [Zt \wedge \neg Zy \longrightarrow Xt]] \wedge$$
$$\wedge\ (\forall x)\ \{Zx\} = \sup^x Z\ \wedge$$
$$\wedge \left\{(\exists y)\ [\ (\forall t)\ t \leq y\ \wedge [Xy \vee \neg Zy]] \vee\right.$$
$$\left.\vee\ [LM\ \wedge\ \sup Z = \{F\}]\right\}.$$

Lemma 3.10: Lemma [Bü] 1.5 is derivable from WLO .

Proof: We have to use lemma 3.8 to get the existence of the recursively introduced set variables. The equivalences are then easily proved by induction. Q.e.d.

We have formalized now the first part of the decision procedure of [Bü] by showing: every formula is, up to a prefix of set quantifiers in front, WLO-derivably equivalent to a formula in automata normal form. To get rid of the prefix, we have to derive the negation lemma using deterministic automata. I.e. we have to formalize section [Bü] 4 to show that to any formula in automata normal form there is a $\overline{\mathcal{C}}_0$-derivably equivalent formula in deterministic automata normal form.

4. The completeness of $\mathcal{A}_\omega$

The formalization will be easier to understand if - as in [Bü] - we look at the case ω first, and generalize to arbitrary ordinals later. Thus we will start by going through item 2 of section [Bü] 3, showing that all the steps are derivable in $\mathcal{A}_\omega$.

Let us recall the definition of $\mathcal{A}_\omega$ in front of theorem 2.3*. For $\alpha=\omega$, $\nu=n_1=0$ and $k=l=n=1$. Thus the 1.case applies, and

$$\mathcal{A}_\omega = \mathcal{C}_0 \cup \left\{ \sum_0^1 \wedge LM_1 \wedge \neg S_2 \right\} .$$

Now $\sum_0^1$ is equivalent to $\neg T_{1,1}$, i.e. to

$$(\forall x \in Lm_1)(\exists y \in Lm_2)\ x \leq y .$$

Therefore $\sum_0^1 \wedge \neg S_2$ is equivalent to $\neg S_1$:

Lemma 4.1: The axiom system $\mathcal{A}_\omega$ consists of $\mathcal{C}_0$ together with LM and $\neg(\exists x)\ Lm\ (x)$, i.e.

$$(\forall y)(\exists z)\ y < z ,$$
$$\neg(\exists x \neq 0)(\forall y)^x (\exists z)^x\ y < z .$$

We will use the notation of section [Bü] 3. We have to be careful, however, about the meaning of the symbols in our formal language. $S^{c,C}$ e.g. is not a two-place predicate variable. $b \in S_u^{c,C}v$ is simply an abbreviation for the formula following the semicolon in the defining formula [Bü] 3.(2). Note that this formula contains the free variables X, and, as most of the abbreviations of this section, depends on the transition system Γ. Also the equations [Bü] 3.(4) are not recursive definitions of S_u, but equivalences to be derived:

Lemma 4.2: The formulas [Bü] 3.(4) are derivable in $\mathcal{A}_\omega$.

Proof: The first formula is trivial. Let $b \in F^{c,C}[Xv, S_u v]$ for $u \leq v$. Then there is $a \in C$ s.t.

$$H[Xv,a,b] \wedge a \in S_u^{c,C}v \cup S_u^{c,C-\{a\}}v .$$

If $a \in S_u^{c,C}v$, then there is a run Z s.t.

$$Zu = c \wedge (\forall t)_u^v H[Xt,Zt,Zt'] \wedge Zv = a \wedge \{Zu,\ldots,Zv-1\} = C .$$

By (COMP), define the run Y as

$$Yt = \begin{cases} Zt & ; t \neq v' \\ b & ; t = v' \end{cases}$$

Or, to give the full detail, if $Z = \langle Z_1,\ldots,Z_n \rangle$ and $b = \langle b_1,\ldots,b_n \rangle$ are n-tuples of predicates and truth values resp., define $Y_1,\ldots,Y_n$ by

$$Y_i t \longleftrightarrow [t \neq v' \wedge Z_i t] \vee [t = v' \wedge b_i] ;$$

these Y_i exist by COMP. Then

$$Yu = c \wedge (\forall t)_u^{v'} H[Xt, Yt, Yt'] \wedge Yv' = b \wedge \{Yu,\ldots,Yv\} = C .$$

Therefore $b \in S_u^{c,C}v'$. Similarly in case $a \in S_u^{c,C-\{a\}}v$. The converse is trivial. Q.e.d.

Again $u \approx v\ (t)$ and $u \approx v\ (-\omega)$ are formulas, containing (besides X) the free variables u, v, t and u, v resp.

<u>Lemma 4.3:</u> Remark [Bü] 3.2 is derivable in A_ω.

<u>Proof:</u> To show that $\approx(t)$ is of index $\leq g$ on t', we have to derive

$$(\exists z_1,\ldots,z_g)^{t'}(\forall u)^{t'} \bigvee_{i=1}^{g} u \approx z_i\ (t) .$$

To this end let $s_1,\ldots,s_g$ be the possible states of S. Then

$$(\forall u)^{t'} \bigvee_{i=1}^{g} S_u t = s_i$$

and

$$(\exists z_1,\ldots,z_g)^{t'} \bigwedge_{i=1}^{g} \left[S_{z_i} t = s_i \vee (z_i = 0 \wedge (\forall u)^{t'} S_u t \neq s_i)\right]$$

are easily derivable. Together we get

$$(\exists z_1,\dots,z_g)^{t'}(\forall u)^{t'} \bigvee_{i=1}^{g} S_u t = S_{z_i} t \; .$$

The rest of the remark is easy . Q.e.d.

Lemma 4.4: Lemma [Bü] 3.3 and formula [Bü] 3.(7) are derivable in A_ω.

Proof: For the derivation of lemma [Bü] 3.3 one has to use twice the pigeon hole principle, lemma 3.4. The existence of the sets Q and P follows from COMP . The set $P = \{y_i\}$ in 3.(7) is defined by a formal recursion as follows

(1) $Pu \longleftrightarrow Qu \wedge (\forall y \in P)^{u} (\exists t)^{u} y \approx w(t)$.

P exists by the recursion lemma 3.8. To show that P is infinite, assume that there is $v \in P$ s.t. $(\forall u \in P)\; u \leq v$. By (1), $v \in Q$. Thus there is $t \geq v$ s.t. $v \sim w(t)$. Since Q is infinite, there is a smallest $z \in Q$, $z > t$. Using (1) twice we get $z \in P$, contradiction. The converse of [Bü] 3.(7) is easy. Q.e.d.

In lemma [Bü] 3.4 we enter the central difficulty of section [Bü] 3 what regards derivations. Here we will use the splicing axiom, indicating afterwards how to eliminate it. (We are in the case ω!) The beginnig of the proof is easy. If P does not satisfy (c), we select an infinite subset by recursion. Similarly for (d) with COMP . For the converse let s_o, s, c, C, d, D, P, y_o be given as in [Bü] , i.e.

(a) $Py \longrightarrow S_o y = s_o$,

(b) $v < y \wedge Pv \wedge Py \longrightarrow S_v y = s$,

(e) $(\exists^{\omega} y)\, Py$,

(f) $Py_o \wedge (\forall t)^{y_o} \neg Pt$,

(g) $E[c]$, $\mathcal{H}[D]$, $d \in s_o^{c,C}$, $d \in s^{d,D}$.

Then we get as there:

(h) $(\exists Z).\ Zo = c \wedge (\forall t)_{o}^{y_o} H[Xt, Zt, Zt'] \wedge Zy_o = d$,

(i) $u < y \wedge Pu \wedge Py \longrightarrow$
$\longrightarrow (\exists Z).\ Zu = d \wedge (\forall t)_{u}^{y} H[Xt, Zt, Zt'] \wedge Zy = d \wedge$
$\wedge \{Zt;\ u \leq t < y\} = D$.

We call the elements of P splicing points. Then (h),(i),(e), and (g) assert that there exists a partial run between any two splicing points in such a way that, if we splice these partial runs together, we get a total run Z for Γ . The problem is that the partial runs are not uniquely determined. Therefore we need the axiom of choice to show that the total run Z exists. For this reason we introduced the splicing axiom at the end of section 2. Thus we define $\phi(y,z,Z)$ as

$$(\exists v) \bigvee_{b \in D} \Big[v' = z \wedge Zy = d \wedge (\forall t)_{y}^{v} H[Xt, Zt, Zt'] \wedge \wedge Zv = b \wedge H[Xv, b, d] \wedge \{Zt;\ y \leq t < z\} = D \Big] .$$

Obviously

(1) $y < z \wedge Y \stackrel{z}{\underset{y}{\equiv}} Z \longrightarrow [\phi(y,z,Y) \longleftrightarrow \phi(y,z,Z)]$.

Further for $y < z$

(2)
$$\Big[Zy = d \wedge (\forall t)_{y}^{z} H[Xt, Zt, Zt'] \wedge Zz = d \wedge \{Zt;\ y \leq t < z\} = D \Big] \longleftrightarrow \longleftrightarrow [\phi(y,z,Z) \wedge Zz = d] ,$$

which yields by (i)

(3) $y < z \wedge Py \wedge Pz \longrightarrow (\exists Z)\ \phi(y,z,Z)$.

By (1) and (3) we get from SPLICE that there is a Z s.t.

$$(\forall y,z \in P)\ [(z = y')\ [P] \longrightarrow \phi(y,z,Z)] .$$

Using (2), (e), and (f), we derive

(4) $Zy_o = d \wedge (\forall t)_{y_o} H[Xt, Zt, Zt'] \wedge \sup Z = D$.

By (h) we can correct the initial partion of Z to get a total run for Γ . In formulas:

$$\tilde{Z}o = c \wedge (\forall t)_o^{y_o} H[Xt,\tilde{Z}t,\tilde{Z}t'] \wedge \tilde{Z}y_o = d \wedge$$

$$\wedge Zy_o = d \wedge (\forall t)_{y_o} H[Xt,Zt,Zt'] \wedge \sup Z = D \wedge$$

$$\wedge (\forall t)_o^{y_o} Yt = \tilde{Z}t \wedge (\forall t)_{y_o} Yt = Zt \longrightarrow \Gamma(X,Y)$$

by (g). If we put existential quantifiers $(\exists \tilde{Z})$, $(\exists Z)$, $(\exists Y)$ in front of the antecedent, and $(\exists Y)$ in front of the succedent, we get $(\exists Y)\ \Gamma(X,Y)$ from (u) , (4) and COMP .

<u>Lemma 4.5:</u> Lemma [Bü] 3.4 is derivable from $\bar{\mathcal{A}}_\omega$.

To get the same result for $\mathcal{A}_\omega$ we will indicate the proof of Siefkes [Si 1] , section I.5.6 , that SPLICE is derivable from $\mathcal{A}_\omega$. Thus let the formula $\phi(y,z,Z^k)$ satisfy

(1) $\quad y < z \wedge Y \overset{z}{\underset{y}{\equiv}} Z \longrightarrow \left[\phi(y,z,Y) \longleftrightarrow \phi(y,z,Z)\right]$,

(2) $\quad (\forall y,z \in U) \left[(z = y')[U] \longrightarrow (\exists Z)\ \ \phi(y,z,Z)\right]$.

We want to derive from $\mathcal{A}_\omega$

(3) $\quad (\exists Z)\ (\forall y,z \in U) \left[(z = y')[U] \longrightarrow \phi(y,z,Z)\right]$.

The idea of [Si 1] is to order the pieces $Z[y,z)$ lexicographically and to define the global Z of (3) by recursion; in each U-intervall $[y,z)$ the minimal $Z[y,z)$ satisfying ϕ is picked out. Thus let the 2^k states of $Z^k = (Z_1,\dots,Z_k)$ be ordered by $<$, define a total ordering $\overset{z}{\underset{y}{<}}$ as

$$Y \overset{z}{\underset{y}{<}} Z \ =_{df}\ (\exists t)_y^z \left[Y \overset{t}{\underset{y}{\equiv}} Z \wedge Yt < Zt \right],$$

$$Y \overset{z}{\underset{y}{\leqq}} Z \ =_{df}\ Y \overset{z}{\underset{y}{\equiv}} Z \vee Y \overset{z}{\underset{y}{<}} Z .$$

From the above formula ϕ we define a formula $M(\phi)$ as

$$\phi(y,z,Z) \wedge (\forall Y) \left[\phi(y,z,Y) \longrightarrow Z \overset{z}{\underset{y}{\leqq}} Y\right].$$

Read $M(\phi)(y,z,Z)$ as "in the interval $[y,z)$, Z is a minimal string satisfying ϕ " . Obviously, without using (1) - (3)

(4) $M(\phi)(y,z,Y) \wedge M(\phi)(y,z,Z) \longrightarrow Y \overset{z}{\underset{y}{=}} Z$,

(5) $M(\phi)(y,z,Z) \longrightarrow \phi(y,z,Z)$,

(6) $(\exists Z)\ \phi(y,z,Z) \longrightarrow (\exists Z)\ M(\phi)(y,z,Z)$.

Formulas (4) and (5) are obviously derivable from WLO . It is not hard but somewhat tedious to derive (6) from A_ω ; details can be found in [Si 1] , section I.5.b , theorems 1 + 2 .

<u>Theorem 4.1</u> (Siefkes [Si 1], section I.5.b): SPLICE is derivable from A_ω.

<u>Proof:</u> Let $\phi(y,z,Z^k)$ satisfy (1) and (2) above. Define $Y = (Y_1,\dots,Y_k)$ by

(7) $$Y_i u \longleftrightarrow (\exists y,z \in U)\left\{y \leq u < z \wedge (z = y')[U] \wedge \wedge (\exists Z)\left[M(\phi)(y,z,Z) \wedge Z_i u\right]\right\} .$$

It is easy to derive from (4) and (7)

$$(z = y')[U] \wedge M(\phi)(y,z,Z) \longrightarrow Z \overset{z}{\underset{y}{=}} Y .$$

By (2) and (1) this yields

$$(z = y')[U] \longrightarrow \phi(y,z,Y) .$$

This proves (3) , since Y exists by COMP . Q.e.d.

<u>Corollary 4.1:</u> Lemma [Bü] 3.4 is derivable from A_ω.

The formula $M(\phi)$ chooses for any y,z a minimal $Z[y,z)$ satisfying $\phi(y,z,Z)$ if such Z exists. More general, we call $\Delta(Z^k)$ a <u>choice formula for the formula</u> $\phi(Z^k)$ if the following holds

(8) $\Delta(Y) \wedge \Delta(Z) \longrightarrow Y = Z$,

(9) $\Delta(Z) \longrightarrow \phi(Z)$,

(10) $(\exists Z)\ \phi(Z) \longrightarrow (\exists Z)\ \Delta(Z)$.

Note that if ϕ contains the free variables $y_1,\dots,y_n$ and $Y_1,\dots,Y_m$ besides Z, then given any $y_1,\dots,y_n$, $Y_1,\dots,Y_m$, Δ picks out a unique Z satisfying $\phi(y_1,\dots,y_n, Y_1,\dots,Y_m, Z)$ if there is such Z. We say that a theory $\mathcal{T}$ satisfies the <u>axiom of definable choice</u> iff to any formula ϕ there is a choice formula Δ such that (8) - (10) are true in $\mathcal{T}$.

<u>Theorem 4.2</u> (Siefkes [Si 2], section 4) : MT [ω] satisfies the axiom of definable choice.

We will sketch the proof, which is an easy generalization of the proof of theorem 4.1. Let the formula $\Psi(X,Y)$ be given, containing no other free variables. We will define a choice formula for $\Psi(Y)$. We assume Ψ in automata normal form $(\exists\tilde{Z})\ \Gamma(X,Y,\tilde{Z})$ where $\Gamma(X,Y,\tilde{Z})$ is

$$E[\tilde{Z}o] \wedge (\forall t)\ H[Xt,Yt,\tilde{Z}t,\tilde{Z}t'] \wedge \mathcal{K}[\sup \tilde{Z}] .$$

Now we consider Y as run instead of as imput; i.e. we define $Z = \langle Y,\tilde{Z} \rangle$ and rewrite Γ as

$$E[Zo] \wedge (\forall t)\ H[Xt,Zt,Zt'] \wedge \mathcal{K}[\sup Z] .$$

We apply the construction of the proof of lemma 4.5 to Γ. Thus let s_o,s be s.t. $\Omega_{s_o,s} \wedge B[s_o,s]$. We define P satisfying (a) - (c) by recursion (see the proof of lemma 4.4). Let $\Delta_{s_o,s}(P,X)$ be a formula describing this recursion. Choose c,D,d,D minimal satisfying (g). Define the formula $\phi_{s_o,s}(y,z,Z,X)$ (somewhat easier than the corresponding formula ϕ in the proof of lemma 4.5) as

$$Zy = d \wedge (\forall t)_y^z\ H[Xt,Zt,Zt'] \wedge Zz = d \wedge \{Zt\ ;\ y \leq t < z\} = D$$

Similarly $\Psi_{s_o,s}(y,z,Z,X)$ as

$$E[Zy] \wedge (\forall t)_y^z\ H[Xt,Zt,Zt'] \wedge Zz = d .$$

Now we define a unique run Z for Γ by the following formula $\chi_{s_o,s}(P,X,Z)$ (cf. formula (7) in the proof of theorem 4.1):

$$\bigwedge_i (\forall u) \Big\{ Z_i u \longleftrightarrow (\exists y,z \in P) \big[y \leq u < z \wedge (z = y') [P] \wedge$$
$$\wedge (\exists \hat{Z}) (M (\Phi_{s_o,s}) (y,z,\hat{Z}) \wedge Z_i u) \big] \vee$$
$$\vee (\exists z \in P) \big[(\forall y \in P)\, z \leq y \wedge (\exists \hat{Z}) (M (\Psi_{s_o,s}) (o,z,\hat{Z}) \wedge \hat{Z}_i u) \big] \Big\} .$$

Define a formula $\Delta(Z,X)$ as

$$\bigvee_{\substack{\{s_o,s;\ B\ [s_o,s]\} \\ s_o,s \text{ minimal}}} (\exists P) \left[\Delta_{s_o,s}(P,X) \wedge \chi_{s_o,s}(P,X,Z) \right].$$

We claim that $\Delta(Z)$ is a choice formula for $\Gamma(Z)$. Condition (8) is obvious. From $\Delta(z)$ we get $\Gamma(Z)$ by formula [Bü] 3.(9), since $(\exists P)\Delta_{s_o,s}(P)$ is equivalent to $\Omega_{s_o,s}$. For the same reason we get $(\exists Z)\ \Delta(Z)$ from $(\exists Z)\ \Gamma(Z)$. Obviously then $(\exists \tilde{Z})\ \Delta(Y,\tilde{Z})$ is a choice formula for $\Psi(Y)$. If Ψ contains free individual variables, we eliminate them using set variables for singletons; see [Si 2] .

From choice formulas for $MT[\omega]$ it is easy to define choice formulas for all $MT[\omega^k]$ by induction on k . Thus we get

<u>Theorem 4.3:</u> For all $\alpha < \omega^\omega$, $MT[\alpha]$ satisfies the axiom of definable choice.

<u>Corollary 4.2:</u> SPLICE is derivable from $\mathcal{A}_\alpha$ for $\alpha < \omega^\omega$.

The construction of theorem 4.3 does not extend to ω^ω, since the choice formulas for $MT[\omega^k]$ depend on k .

<u>Problem:</u> Does (a) $MT[\omega^\omega]$, (b) $MT[co]$ admit choice formulas ?

<u>Lemma 4.6:</u> The normal form theorems [Bü] 3.7+8 hold for $\mathcal{A}_\omega$. I.e. to any formula $\Sigma(X)$ there is a formula in automata normal form,

$$(\exists Z).\ A[Zo] \wedge (\forall t)\ B\ [Xt,Zt,Zt'] \wedge (\exists^\omega t)\ C\ [Zt] ,$$

which is $\mathcal{A}_\omega$-derivably equivalent to $\Sigma(X)$.

<u>Proof:</u> If in the definition of E' and H' in the proof of lemma

[Bü] 3.5 we cancel the condition

(1) $\quad Pt' \wedge Qt' \longrightarrow F[Xt, Y_2t] = Y_1t'$,

and add e.g :

$$\neg Qt \wedge \neg Pt \longrightarrow Y_1t = Y_2t = e \ ,$$

the existence of Q, Y_0, Y_1, Y_2 follows from the recursion lemma 3.8. Then from the P of $\Omega_{s_0,s}$ we get $(1) \wedge \phi_{s_0,s}(P, Y_0, Y_1)$ where $\phi_{s_0,s}$ is the terminal condition of [Bü] . Conversely we get $P_1 \subseteq P$ from $(1) \wedge \phi_{s_0 s}(P, Y_0, Y_1)$ by comprehension. This derives lemma [Bü] 3.5 . For the complementation lemma [Bü] 3.6 we need the pigeon hole lemma 3.4 .

Q.e.d.

Lemma 4.7: Remark [Bü] 3.9 is derivable from A_ω.

Proof: Note that the obvious equations (a) and (b) of the proof in [Bü] are not expressible in the language of MT [co] . It is trivial, however, to derive from A_ω

(a') $\quad S_0 n = S_u(u+n)$,

(b') $\quad S_0 x = S_0 y \longrightarrow [S_0(x+n) = S_0(y+n)]$

for every n . We just have to use the recursion [Bü] 3.(4) , which is derivable by lemma 4.2 . Let g be the number of states of S_0 . Then there are $m < m+n \leq g$ s.t.

$$A_\omega \vdash S_0 m = S_0(m+n) \ .$$

As in [Bü] we get from (a') by induction on y that

(c') $\quad (\forall y)[y \equiv 0(n) \longrightarrow y \approx 0\ (-\omega)]$

is derivable from A_ω . Again as in [Bü] there are $i < i+j \leq g$ and a state s s.t.

$$A_\omega \vdash S_0 v i = S_0 v(i+j) = s \ .$$

By induction on y we get from (b') and (c')

$$A_\omega \vdash (\forall y)[y \equiv ni\,(nj) \longrightarrow y \approx 0\ (-\omega) \wedge S_0 y = s] ,$$

thus $\mathbb{A}_\omega \vdash \Omega_{s,s}$. (See section 2 for the definition of congruence)

Q.e.d.

Lemmata 4.6+7 yield

<u>Theorem 4.4</u> (Siefkes [Si 1] , theorem I.5.a.1): $\mathbb{A}_\omega$ is a complete axiom system for MT[ω] .

5. The deterministic automata normal form

We will now formalize section [Bü] 4 . This will show that the deterministic automata normal form can be derived from $\overline{\mathcal{C}_0}$. Since we have treated the case ω separately, the beginning (including formula [Bü] 4.(4)) is easy. Although the construction of the deterministic automata for MT [co] is rather difficult, the derivations involved are not hard. (Therefore we omitted the formalization of item 2' in section [Bü] 3.) We will thus go through section [Bü] 4 , giving (hints for) derivations where necessary.

We work with the transition system $\Gamma(X,Z)$ of [Bü] 4.(1). For the notations $b \in S_u^{c,C} v$ and $u \approx v(-x)$ see the beginnig of section 4 . A derivation for lemma [Bü] 4.1 can be taken from lemma 4.2 . The first part of remark [Bü] 4.2 follows from lemma 4.3 by the pigeon hole lemma 3.4 . To get the second part, let $w < x$ and Q satisfy

(1) $\quad \mathrm{Cof}\,(Q)\,[-x]$,

(2) $\quad (\forall u \in Q)\ u \approx w\ (-x)$.

As in the proof of lemma 4.4 we define P by recursion:

(3) $\quad (\forall u)\ \left[Pu \longleftrightarrow Qu \wedge (\forall y \in P)^u\ (\exists t)^u\ y \approx w\ (t)\right]$.

We want to show

(4) $\quad \mathrm{Cof}\,(P)\ [Q]$.

Thus let $y \in Q$, $y \notin P$. Assume

(5) $\quad (\forall u \in P)\ u < y$.

The assumption

$$(\exists v \in P)^y\ (\forall u \in P)^y\ u \leqslant v$$

leads to a contradiction as in the proof of lemma 4.4 . Therefore

(6) $\quad (\forall v \in P)^y\ (\exists u \in P)^y\ v < u$.

Let $v \in P$, $v < y$. By (6) there is $u \in P$, $u > v$. By (3),

$(\exists t)^u\ v \approx w\ (t)$. This proves

$$(\forall v \in P)^y\ (\exists t)^y\ v \approx w\ (t)\ ,$$

which yields $y \in P$ by (3); contradiction. This shows that (5) is wrong, which proves (4). By (3),

$$(7) \qquad Pu \wedge Pv \wedge (v = u')\,[P] \longrightarrow u \approx w\ (v)\ .$$

From theorem 2.2 we get ACC $[P]$. Thus P contains a cofinal w-sequence. This shows that remark [Bü] 4.2 is derivable from C_o .

If in the proof of the equivalence [Bü] 4.(4) we replace the y_i by the above P , we can translate the proof directly into a derivation; see lemmata 4.4+ 5 . Also the application of the splicing axiom is the same as in lemma 4.5 , except that we have to take care of the limit numbers in P . Thus let $\phi(y,z,Z)$ be the formula

$$Zy = d \wedge (\forall v)_y^z\ (\forall t)_y^v\ H\,[Xt, Zt, Zt'] \wedge (\forall \bar{x})_y^z\ \mathcal{H}\,[\sup{}^{\bar{x}} Z, Z\bar{x}] \wedge$$
$$\wedge \{Zt;\ y \leq t < z\} = D \wedge \Big\{ [Lm(z) \wedge \mathcal{H}\,[\sup{}^z Z,\ d]] \vee$$
$$\vee [(\exists v) \bigvee_{b \in D} [z = v' \wedge Zv = b \wedge H\,[Xt,b,d]]] \Big\}\ .$$

If we apply SPLICE to ϕ , we get the run $Z\,[u,x]$ as in lemma 4.5 .

Now we will formalize the construction of deterministic automata normal forms, which begins with formula [Bü] 4.(5) . The first half of this equivalence is derived in the same way as remark [Bü] 4.2 above. For the second half of the equivalence define the set T by

$$Tt \longleftrightarrow (\exists y)_{u'}^x\ [y \sim w\ (t) \wedge b \in B(y)]\ .$$

Then $(\exists^x t)\,Tt$. Since for any y , S_y has only g states, we get

$$\bigwedge_{i=1}^{2} [y_i \leq z < t_i \wedge y_i \sim w\ (t_i)] \wedge t_1 \neq t_2 \longrightarrow S_{y_1} z = S_{y_2} z\ .$$

From $(\exists^x t)\,Tt$ follows for $z < x$

$$(\exists t_o, \ldots, t_g \in T)_z \bigwedge_{i \neq j} t_i \neq t_j\ .$$

Together we get

$$(\forall z)^x\ (\exists y)_z\ (\exists t)\ [y \sim w\ (t) \wedge b \in B(y)]\ .$$

The expressions in (6) are to be read as abbreviations for the following formulae:

$$r_u t = z \equiv_{df} z \sim u\,(-t) \wedge (\forall v)^{z}\ v \not\sim u\,(-t)\ ,$$

$$v \in Ut \equiv_{df} (\exists u)^{t}\ v = r_u\, t\ ,$$

$$u \longleftarrow v(t) \equiv_{df} v \in Ut \wedge u \in Ut \wedge u \sim v\ (t) \wedge (\forall w \in Ut)^{u}\ w \not\sim v\ (t)\ ,$$

$$\langle b,\ r_y\ t \rangle \in Q^{c,C}_{u,w}\ t \equiv_{df} u,w < y < t \wedge b \in B^{c,C}[S_u y\ ,\ S_w y]\ .$$

The proofs for [Bü] 4.(7) and lemmma [Bü] 4.3 are then easily translated into derivations.

In [Bü] 4.(8) S_i , V and $Q_{i,k}$ are finite state, and can thus be coded by set variables; $\mathcal{S}_i$ and $\mathcal{Q}_{i,k}$ are used for abbreviations. The abbreviations $i \longleftarrow j\,(t)$ and $\rho(t) = i$ are easily translated into formulas containing only the free variable t and no quantifiers. For the proof of lemma [Bü] 4.4 note that the recursions of [Bü]4.($\bar{8}$) are not, like in (8), finite state. We thus have to define $\bar{S}, \bar{Q}, U$ as in [Bü] 4.(2)+(6), and to derive 4.($\bar{8}$) for the appropriate abbreviations. We further define

$$u_i\ t = v \equiv_{df} \dot{v} < t \wedge i = \rho(v) \wedge (\forall y)^{t}_{v'}\ i \longleftarrow i(y)\ ,$$

and derive the equations (c) in the form

$$i = \rho(t) \longrightarrow u_i t' = t\ ,\ i \longleftarrow i(t) \wedge u_i t = v \longrightarrow u_i t' = v\ ,$$

$$(\forall^{x} t\ )\ [\, i \longleftarrow i(t) \wedge u_i\, t = v \,] \longrightarrow u_i\, x = v\ .$$

The proofs for (b_1) - (b_6) yield derivations directly.

Thus we have put together a derivation for lemma [Bü] 4.4 , if we provide the right terminal condition. Recall that we are looking for a formula in deterministic automata normal form which is derivably equivalent to a formula of the form

$$(\exists Z)\ .\ \mathcal{E}[Zo] \wedge (\forall t)\ H\,[Xt, Zt, Zt'] \wedge (\forall x)\ \mathcal{K}[\sup{}^{x} Z, Zx] \wedge \mathcal{L}\,[\sup X,\ \sup Z]\ ,$$

where $\mathcal{L}[\sup X, \sup Z]$ is short for

$$(\exists y)\left[(\forall t)\ t \leq y \wedge L_0[Xy, Zy]\right] \vee \left[LM \wedge \mathcal{L}_1[\sup Z]\right].$$

As in section [Bü] 3 we write

$$\Gamma(X,Z) \equiv_{df} E[Zo] \wedge (\forall t)\ H[Xt, Zt, Zt'] \wedge (\forall x)\ \mathcal{H}[\sup^x Z, Zx].$$

Similarly we use $\Gamma'(X,Y)$ for the conjunction of the recursions [Bü] 4.(8); thus Y stands for all the $S_i, V, Q_{i,k}$. Then we have to find a terminal condition $\mathcal{L}'(X,Y)$ of the same form as $\mathcal{L}$ such that

$$(\exists Z)\left[\Gamma(X,Z) \wedge \mathcal{L}[\sup X, \sup Z]\right] \longleftrightarrow$$
$$\longleftrightarrow (\exists Y)\left[\Gamma'(X,Y) \wedge \mathcal{L}'[\sup X, \sup Z]\right].$$

Lemma [Bü] 4.4 provides us with an $\mathcal{L}'$ which works in case of successor type structures. We have to take care of the limit case as well. Obviously it is enough if we split the above equivalence and prove:

$$(\exists y)\left[(\forall t)\ t \leq y \wedge \left\{(\exists Z)\left[\Gamma(X,Z) \wedge L_0[Xy, Zy]\right] \longleftrightarrow \right.\right.$$
$$\left.\left.\longleftrightarrow (\exists Y)\left[\Gamma'(X,Y) \wedge L_0'[Xy, Yy]\right]\right\}\right] \vee$$
$$\vee \left[LM \wedge \left\{(\exists Z)\left[\Gamma(X,Z) \wedge \mathcal{L}_1[\sup Z]\right] \longleftrightarrow \right.\right.$$
$$\left.\left.\longleftrightarrow (\exists Y)\left[\Gamma'(X,Y) \wedge \mathcal{L}_1'[\sup Y]\right]\right\}\right]$$

for suitable L_0', $\mathcal{L}_1'$. For L_0' we let the terminal condition of lemma [Bü] 4.4 depend on X:

$$L_0'[Xt, Yt] \equiv_{df} \bigvee_{c,C,b} .\ E[c] \wedge b \in S_0^{c,C}t \wedge L_0[Xt, b].$$

For $\mathcal{L}_1'$ we use the construction of section [Bü] 3.

Define

$$B'[s_0, s] \equiv_{df} \bigvee_{c,C,d,D} .\ E[c] \wedge \mathcal{L}_1[D] \wedge d \in s_0^{c,C} \wedge s^{d,D}.$$

Then we get as there, under the premise LM:

$$(\exists Z)\left\{\Gamma(X,Z) \wedge \mathcal{L}_1 [\sup Z]\right\} \longleftrightarrow$$
$$\longleftrightarrow (\exists w)\ (\exists^{\infty} y)\ \left\{y \sim w\ (-\infty) \wedge B\left[S_o y\ ,\ S_w y\right]\right\} \longleftrightarrow$$
$$\longleftrightarrow (\exists w)\left\{(\forall^{\infty} t)\ w \longleftarrow w(t) \wedge (\exists^{\infty} t)\ (\exists v)\left[v \neq w \longleftarrow v(t) \wedge \right.\right.$$
$$\left.\left. \wedge\ v \in Q_w t\right]\right\},$$

where

$$Q_w t =_{df} \left\{r_y t\ ;\ w < y < t \wedge B\left[S_o y\ ,\ S_w y\right]\right\} .$$

Here we have used the unrestricted cofinality quantifiers

$$(\exists^{\infty} t)\ \phi(t) \equiv_{df} (\forall y)\ (\exists t)_y\ \phi(t),\ \text{i.e. } \mathrm{Cof}\ (\{t\ ;\ \phi(t)\})\ ,$$
$$(\forall^{\infty} t)\ \phi(t) \equiv_{df} (\exists y)\ (\forall t)_y\ \phi(t),\ \text{i.e. } \neg\,\mathrm{Cof}\ (\{t\ ;\ \neg\phi(t)\})\ .$$

As in lemmata [Bü] 3.12 and [Bü] 4.3 we prove the recursion equations

$$Q_w w = \emptyset\ ,$$
$$Q_w t' = \left\{v\ ;\ \bigvee_p \left[v \longleftarrow p(t) \wedge p \in Q_w t\right] \vee \left[v = t \wedge B\left[S_o t\ ,\ S_w t\right]\right]\right\},$$
$$Q_w x = \left\{v\ ;\ (\forall^x t)\ v \in Q_w t\right\} .$$

Thus we replace Q by the finite state Q_k , enlarge Γ' by the recursions

$$Q_k o = \emptyset\ ,\ Q_k t' = \begin{cases}\emptyset\ ;\ k \not\longleftarrow k(t) \\ \mathcal{D}_k(t')\ ;\ \text{otherwise}\end{cases},\ Q_u x = \begin{cases}\emptyset\ ;\ (\exists^x t)\ k \longleftarrow k(t) \\ \mathcal{D}_k\ (x)\ ;\ \text{otherwise}\end{cases}$$

where

$$\mathcal{D}_k(t') =_{df} \left\{j\ ;\ \bigvee_h \left[j \longleftarrow h(t) \wedge h \in Q_k t\right] \vee \left[j = \rho(t) \wedge B\left[S_o t, S_k t\right]\right]\right\}$$
$$\mathcal{D}_k\ (x) =_{df} \left\{j\ ;\ (\forall^x t)\ j \in Q_k t\right\},$$

and define

$$\mathcal{L}_1'\ [\sup Y] \equiv_{df} \bigvee_k\ .\ (\forall^{\infty} t)\ k \longleftarrow k(t)\ \wedge$$
$$\wedge\ (\exists^{\infty} t)\ (\exists v)\left[v \neq w \longleftarrow v(t) \wedge v \in Q_w t\right].$$

For these formulae L_o' , $\mathcal{L}_1'$ we derive the wanted equivalence. Thus we have proved:

Lemma 5.1 : Lemma [Bü] 4.4 is derivable from $\bar{\mathscr{C}}_0$: To any formula in automata normal form there is a $\bar{\mathscr{C}}_0$-derivably equivalent formula in deterministic automata normal form.

Theorem 5.1 : Theorem Bü 4.5 is derivable: To any formula $\Sigma(X)$ there is a formula $\Phi(X)$ in deterministic automata normal form,

$$(\exists Y) \,.\, Y_0 = a \wedge (\forall t)\ Yt' = F[Xt, Yt] \wedge \wedge (\forall x)\ Yx = \mathscr{G}[\sup{}^{x} Y] \wedge \mathscr{L}[\sup X, \sup Z],$$

such that

$$\bar{\mathscr{C}}_0 \vdash \Sigma(X) \longleftrightarrow \Phi(X) .$$

6. $\bar{\mathscr{C}}_0$ and $\bar{A}_\alpha$, $\alpha<\omega_1$, are axiom systems for MT [co] and MT[α]

Now that we have derived from $\bar{\mathscr{C}}_0$ the deterministic automata normal form for MT [co] formulas, we have to use the decidability of sentences in this form to show that any sentence in this form which belongs to MT [co], is derivable from $\mathscr{C}_0$. Throughout this section "derivable" will mean "derivable from $\mathscr{C}_0$". Let ϕ be a sentence in deterministic automata normal form:

(1) $$(\exists Z) \,.\, Zo = a \wedge (\forall t)\, Zt' = F[Zt] \wedge (\forall x)\, Zx = \mathcal{G}[\sup{}^x Z] \wedge \wedge \mathcal{L}[\sup Z] .$$

We will write

(2) $$\Gamma(Z) \equiv_{df} (\forall t)\, Zt' = F[Zt] \wedge (\forall x)\, Zx = \mathcal{G}[\sup{}^x Z] ,$$
$$\Gamma_c(Z) \equiv_{df} Zo = c \wedge \Gamma(Z) \quad \text{for any state } c .$$

Let K be the set of states of the automaton given by (1). K is finite, and $F : K \longrightarrow K$, $\mathcal{G} : \mathcal{P}(K) \longrightarrow K$. As in lemma [Bü] 4.11 define mappings Z_c for $c \in K$, F_i, V_i, S_{i+1} for $i < \omega$ as follows:

(3) $$Z_c o = c \,,\; Z_c t' = F[Z_c t] \,,\; Z_c x = \mathcal{G}[\sup{}^x Z_c] ,$$
$$F_i\, [c] =_{df} Z_c\, \omega^i ,$$
$$V_i\, [c] =_{df} \{ Z_c t \,;\, t < \omega^i \} ,$$
$$S_i\, [c] =_{df} \sup{}^{\omega^i} Z_c .$$

Then $Z_c : \omega_1 \longrightarrow K$, $F_i : K \longrightarrow K$, $V_i, S_i : K \longrightarrow \mathcal{P}(K)$. From (1) we get

<u>Corollary 6.1 :</u> $\vdash \phi \longleftrightarrow \mathcal{L}[\sup Z_c]$, or more formally:

$$\vdash \Gamma_c\,(Z) \longrightarrow [\phi \longleftrightarrow \mathcal{L}[\sup Z]] .$$

By corollary 2.3, ω^i is definable in any model where it exists. Therefore the definitions of F_i etc. are expressible in the language. E.g. $F_i\, [c] = d$ is short for

$$(\forall v)\, (\forall Z) \,.\, \Sigma_{\omega^i}\, [v] \wedge Zo = c \wedge (\forall t)\, Zt' = F[Zt] \wedge \wedge (\forall x)\, Zx = \mathcal{G}[\sup{}^x Z] \longrightarrow Zv = d .$$

Thus the recursions of lemma [Bü] 4.11 are expressible, too. They are, however, rather hard to derive. Consider for example the equation

$$F_1^{\,2}[c] = Z_c\,\omega^2 \ , \text{ i.e. } Z_{Z_c\omega}\,\omega = Z_c\,\omega^2 \ .$$

This shows that we have to derive that the value of a run Z at time $t+\alpha$ depends only on Zt, but not on t itself. I.e. we have to derive (cf. the proof of lemma 4.7)

$$Z_c v = Z_d u \longrightarrow Z_c v + \alpha = Z_d u + \alpha \ ,$$

or, what amounts to the same,

$$(4) \qquad Z_c v = d \longrightarrow Z_c v + \alpha = Z_d \alpha \ .$$

Note that these statements are expressible iff α is definable. In fact, if $\alpha < \omega^\omega$, by corollary 2.3 we can write

$$w = v + \alpha \quad \text{short for} \quad \textstyle\sum_\alpha\{v\}\,[w] \ ,$$

and get as a more formal expression

$$Z_c v = d \ \wedge\ \textstyle\sum_\alpha\{v\}\,[w] \ \wedge\ \sum_\alpha[u] \longrightarrow Z_c w = Z_d u .$$

To prove (4) note that before relativizing a formula one has to eliminate all defined concepts. E.g. $Zo = d$ has to be written as

$$(\forall y)\ \big[(\forall t)\ y \leq t \longrightarrow Zy = d\big] .$$

This shows that

$$(Zo = d)\ \{v\} \longleftrightarrow Zv = d \ .$$

Further $Z_d\,\alpha = e$ is short for

$$(\forall Z)\ (\forall w)\ \big[\Gamma_d\,(Z)\ \wedge\ \textstyle\sum_\alpha[w] \longrightarrow Zw = e\big] .$$

Thus

$$(Z_d\,\alpha = e)\ \{v\} \longleftrightarrow (\forall Z)\ (\forall w)_v\ \big[Zv = d\ \wedge\ (\forall t)_v\ Zt' = F\,[Zt]\ \wedge$$
$$(\forall x)_v\ Zx = \mathcal{G}[\sup{}^x Z]\ \wedge\ \textstyle\sum_\alpha\{v\}\,[w] \longrightarrow Zw = e\big] .$$

Therefore

$$(Z_d\,\alpha = e)\ \{v\} \longrightarrow (\forall Z)\ (\forall w)_v\ \big[\Gamma_c(Z)\ \wedge Zv = d\ \wedge\ \textstyle\sum_\alpha\{v\}\,[w] \longrightarrow$$
$$\longrightarrow Zw = e\big] \ ,$$

i.e.

$$(Z_d\alpha = e)\,\{v\} \wedge Z_c v = d \longrightarrow Z_c v + \alpha = e ,$$

i.e.

(5) $Z_d\alpha = e \wedge (Z_d\alpha = e)\,\{v\} \wedge Z_c v = d \longrightarrow Z_c v + \alpha = Z_d\alpha$.

Now assume

(6) there is $e \in k$ s.t. $\vdash Z_d\alpha = e$.

Then by corollary 2.2

$$\vdash (\forall v)\ (Z_d\alpha = e)\ \{v\} ,$$

which by (5) proves that (4) is derivable. Whereas it is easy to show by induction on t that

$$\vdash \bigvee_{e \in k} Z_d t = e ,$$

there seems to be no direct proof for (6). We cannot do induction on α, since for the induction step on limits this would involve infinite derivations. Therefore we have to prove restricted versions of (6) and (4) simultaneously with lemma [Bü] 4.11 ; we will get (4) as a corollary.

<u>Lemma 6.1 :</u> Lemma [Bü] 4.11 is derivable, indeed:

(a) $\vdash F_o\,[c] = F\,[c] \wedge V_o\,[c] = \{c\} = S_o\,[c]$,

(b) for any $d \in k$ there is $e \in k$ s.t. $\vdash Z_d\,\omega^i = e$,

(c) $\vdash Z_c v = d \longrightarrow Z_c v + \omega^i = F_i\,[d]$,

(d) $\vdash Z_c\,\omega^i \cdot j = F_i^j\,[c]$,

(e) $\vdash Z_c v = d \longrightarrow \{Z_c t ;\ v \leq t < v + \omega^{i+1}\} = V_{i+1}[d]$,

(f) there is a first pair $h_i < k_i$ s.t. for all $c \in k : \vdash F_i^{h_i}\,[c] = F_i^{k_i}[c]$,

(g) $\vdash V_{i+1}\,[c] = \bigcup_{j=o}^{k_i-1} V_i\left[F_i^j\,[c]\right]$,

(h) $\vdash S_{i+1}\,[c] = \bigcup_{j=h_i}^{k_i-1} V_i\left[F_i^j\,[c]\right]$,

(i) $\qquad \vdash F_{i+1}[c] = \mathcal{G}[S_{i+1}[c]]$,

(j) $\qquad \vdash Z_c v = d \longrightarrow \sup^{v+\omega^i} Z_c = S_i[d]$.

<u>Proof :</u> (a) and (i) follow directly from (3). We will prove (b) - (h) by simultaneous induction on i . For any i ,(c) follows from (b) by the considerations in front of this lemma. In exactly the same way we get (e) from (g) and (j) from (h) .

<u>Induction beginning: i = o :</u>

(b) is trivial with $e = F[d]$. (d) is easy with induction on j . To prove (f) note that $F^j[c] = d$ is a propositional formula; therefore any true equation of this form is derivable. Thus (f) follows from the finiteness of $\mathbb{K}$.

To get (g) and (h) we prove by induction on t

(1) $\qquad t < \omega \longrightarrow \bigvee_{j=o}^{n-1} t \equiv j\ (n)$.

Again by induction on t , using (d) and (f) we get for any j , $k_o \leq j < h_o$:

(2) $\qquad h_o \leq t < \omega \wedge t \equiv j\ (k_o - h_o) \longrightarrow Z_c t = Z_c j$.

(1) and (2) togehter yield

(3) $\qquad h_o \leq t < \omega \longrightarrow \bigvee_{j=h_o}^{k_o-1} Z_c t = Z_c j$,

and thus

(4) $\qquad \{Z_c t\ ;\ h_o \leq t < \omega\} = \{Z_c t\ ;\ h_o \leq t < k_o\}$.

Therefore, by (4), (d) and (a) :

$$V_1[c] = \{Z_c t\ ;\ t < \omega\} = \{Z_c t\ ;\ t < k_o\} = \bigcup_{j=o}^{k_o-1} \{Z_c j\} = \bigcup_{j=o}^{k_o-1} V_o[F_o^j[c]] ,$$

which proves (g). Also from (4) we get

(5) $\qquad \sup^\omega Z_c \subseteq \{Z_c t\ ;\ h_o \leq t < k_o\}$.

Now let y be s.t. $h_o \leq y < \omega$. By (1), $y = j\,(k_o - h_o)$ for some j where $h_o \leq j < k_o$. By (2)

$$\{Z_c\, y+k \,;\, o \leq k < k_o - h_o\} = \{Z_c\, t \,;\, h_o \leq t < k_o\},$$

which implies

$$\{Z_c\, t \,;\, h_o \leq t < k_o\} \subseteq \sup^{\omega} Z_c ,$$

and thus by (5)

$$\sup^{\omega} Z_c = \{Z_c\, t \,;\, h_o \leq t < k_o\} .$$

From this we get (h) immediately.

Induction step: Assume (b) - (h) to be proved for $i-1$, prove them for $i > o$.

Since $i > o$, $Z_d\, \omega^i = \mathcal{G}[S_i\,[d]]$, which proves (b) using (h) for $i-1$. We get (d) by induction on j, using (c):

$$Z_c\, \omega^i \cdot (j+1) = Z_c(\omega^i \cdot j + \omega^i) =$$

$$= F_i\left[Z_c\, \omega^i \cdot j\right] = F_i\left[F_i^{\,j}\,[c]\right] = F_i^{\,j+1}\,[c] .$$

From (d) we get (f) (using (h) for $i-1$) and

$$(6) \qquad V_i\left[F_i^{\,j}\,[c]\right] = \{Z_c\, t \,;\, \omega^i \cdot j \leq t < \omega^i \cdot (j+1)\}$$

(using (e) for $i-1$). By relativizing (3) and its proof to Im_i we get by corollary 2.1

$$(7) \qquad \omega^i \cdot h_i \leq t < \omega^{i+1} \wedge Im_i(t) \longrightarrow \bigvee_{j=h_i}^{k_i-1} Z_c\, t = Z_c\, \omega^i \cdot j .$$

Now let $\omega^i \cdot k_i \leq y < \omega^{i+1}$. If $y \notin Im_i$, there is a greatest $z \in Im_i$, $z < y$. Thus there is $z \in Im_i$, $\omega^i \cdot k_i \leq z \leq y < z + \omega^i$. By (7) and (d) $Z_c\, z = F_i^{\,j}\,[c]$ for some $h_i \leq j < k_i$. By (e) for $i-1$, $Z_c\, y \in V_i\left[F_i^{\,j}\,[c]\right]$. This yields together with (6):

$$(8) \qquad \{Z_c t \,;\, \omega^i \cdot h_i \leq t < \omega^{i+1}\} = \bigcup_{j=h_i}^{k_i-1} V_i\left[F_i^{\,j}\,[c]\right] .$$

(8) and (6) imply (g), and moreover

$$(9) \qquad \sup^{\omega^{i+1}} Z_c \subseteq \bigcup_{j=h_i}^{k_i-1} V_i\left[F_i\,[c]\right] .$$

Now let $y < \omega^{i+1}$. There is $z \in Lm_i$, $y \leq z < \omega^{i+1}$, $\omega^i \cdot h_i \leq z$. By (7) and (d) $Z_c z = F_i^l [c]$ for some $h_i \leq l < k_i$. Therefore by (6)

$$\{Z_c t \,;\, z \leq t < z + \omega^i \cdot (k_i - h_i)\} = \bigcup_{j=h_i}^{k_i - 1} V_i \left[F_i^j [c]\right].$$

Thus

$$\bigcup_{j=h_i}^{k_i-1} V_i \left[F_i^j [c]\right] \subseteq \sup{}^{\omega^{i+1}} Z_c \,,$$

which implies (h) by (9). Q.e.d.

Corollary 6.2 : Let $\alpha = \sum_{i=k}^{o} \omega^i \cdot n_i < \omega^\omega$.

(a) $\vdash Z_c \alpha = F_o^{n_o} \ldots F_k^{n_k} [c]$,

(b) $\vdash Z_c v = d \longrightarrow Z_c v + \alpha = Z_d \alpha$.

Proof: (a) is immediate from (c) and (d) in lemma 6.1 ; (b) follows from (a) by the considerations in front of lemma 6.1. Q.e.d.

The recursions of lemma [Bü] 4.11 make it easy to derive lemma [Bü] 4.7. We prove a lemma first:

Lemma 6.2 : The following statements are derivable:

(a) $V_i [c] \subseteq V_{i+1} [c]$.

(b) $c \in V_i [c]$.

(c) $F_i [c] \in V_{i+1} [c]$.

(d) If $d \in V_i [c]$, then $F_i [c] = F_i [d]$, $V_i [d] \subseteq V_i [c]$, and $S_i [c] = S_i [d]$.

(e) $F_j F_i = F_j$ for $i < j$.

(f) If $V_i [c] = V_{i+1} [c]$, then $F_i [c] = F_i^2 [c]$ and $S_i [c] \subseteq S_{i+1} [c]$.

Proof: We use lemma 6.1 with the numbers $h_i < k_i$ defined there.

(a) $V_{i+1} [c] = V_i [c] \cup \ldots$

(b) Induction on i, using (a).

(c) 1.case: $k_i = 1$. Then $F_i[c] = c \in V_{i+1}[c]$ by (b).

2.case: $k_i > 1$. Then $V_{i+1}[c] = V_i[c] \cup V_i\left[F_i[c]\right] \cup \ldots$

Thus $F_i[c] \in V_i\left[F_i[c]\right] \subseteq V_{i+1}[c]$ by (b).

(d) Induction on i: If $d \in V_o[c]$, then $d = c$.

So let $d \in V_{i+1}[c]$. Then there is $j < k_i$ such that $d \in V_i\left[F_i^j[c]\right]$.
By induction hypothesis $F_i^{j+1}[c] = F_i[d]$. Therefore

$$S_{i+1}[d] = V_i\left[F_i^{h_i}[d]\right] \cup \ldots \cup V_i\left[F_i^{k_i-1}[d]\right] =$$

$$= V_i\left[F_i^{h_i+j}[c]\right] \cup \ldots \cup V_i\left[F_i^{k_i+j-1}[c]\right] =$$

$$= V_i\left[F_i^{h_i}[c]\right] \cup \ldots \cup V_i\left[F_i^{k_i-1}[c]\right] = S_{i+1}[c],$$

which implies $F_{i+1}[c] = F_{i+1}[d]$.
Similarly, $V_{i+1}[d] =$

$$= V_i\left[F_i^{\min(j,h_i)}[c]\right] \cup \ldots \cup V_i\left[F_i^{k_i-1}[c]\right] \subseteq V_{i+1}[c]$$

(e) By (c) and (a) $F_i[c] \in V_j[c]$ for $i < j$. Thus $F_j[c] =$
$= F_j\left[F_i[c]\right]$ by (d).

(f) Let $V_i[c] = V_{i+1}[c]$. Then by (c) $F_i[c] \in V_i[c]$, thus $F_i[c] = F_i^2[c]$ by (d). Therefore $S_{i+1}[c] = V_i\left[F_i[c]\right]$, by the definition of h_i, k_i. Let $i > o$. Since $V_i[c] = V_{i+1}[c]$, we get $S_i[c] = S_i\left[F_i[c]\right]$ from (c) and (d).
$S_i\left[F_i[c]\right] \subseteq V_i\left[F_i[c]\right]$ by definition. Q.e.d.

<u>Lemma 6.3</u> : Lemma [Bü] 4.7 is derivable: There is a number p s.t.

(a) $F_p^2 = F_p$,

(b) for every $c \in k$: $F_p[c] = \mathcal{G}[D_c]$ and

$D_c = V_p\left[F_p[c]\right] = S_p\left[F_p[c]\right] = S_p[c]$,

where $D_c =_{df} \left\{Z_c t \,;\, \omega^p \leq t < \omega^p \cdot 2\right\}$,

(c) for every $j \geq p$: $F_j = F_p$.

Proof: By lemma 6.1 any equation concerning F_i, V_i, S_i is a propositional formula; therefore any true equation of this kind is derivable. Thus, since K is finite, by (a) and (f) in lemma 6.2 there is a number p s.t. $V_p = V_{p+1}$ and $S_p = S_{p+1}$. Again by (f) in 6.2, $F_p^2 = F_p$, i.e. $k_p = 2$. Further $F_{p+1}[c] = \mathcal{G}\left[S_{p+1}[c]\right] = \mathcal{G}\left[S_p[c]\right] = F_p[c]$, thus $F_{p+1} = F_p$. By induction we get for any $i \geq p$

$$F_i^2 = F_i \ , \ V_{i+1} = V_i \ , \ S_{i+1} = S_i \ , \ F_{i+1} = F_i \ .$$

By (e) in lemma 6.1,

$$D_c =_{df} \left\{ Z_c t \ ; \ \omega^p \leq t < \omega^p \cdot 2 \right\} = V_p\left[F_p[c]\right] = S_{p+1}[c] \ .$$

Therefore $F_p[c] = F_{p+1}[c] = \mathcal{G}[D_c]$. Q.e.d.

Note that lemma [Bü] 4.7 is actually stronger, since it proves (c) for infinite j, too. ω^ν for infinite ν, however, is not definable in MT [co], therefore we cannot define F_ν. It will be enough if we derive the following expressible analogue to the missing part of lemma [Bü] 4.7 :

$$Lm_j(v) \wedge v \neq o \longrightarrow Z_c v = F_p[c] \quad \text{for any } j \geq p \ .$$

See lemma 6.5 below.

Lemma 6.4 : For any j :

$$\vdash lm_j(v) \longrightarrow (\exists x \in Lm_j)^v \ v = x + \omega^j \ .$$

Proof: Let $lm_j(v)$. Then $\neg Lm_{j+1}(v)$, i.e. there is $x < v$, $x \in Lm_j$, $(\forall t \in Lm_j)^v \ t \leq x$. It is easy to derive $\sum_\omega j\{x\}[v]$, i.e. $v = x + \omega^j$. Q.e.d.

Lemma 6.5 : Let p be given from lemma 6.3, let $j \geq p$:

$$\vdash Lm_j(v) \wedge v \neq o \longrightarrow Z_c v = F_p[c] \ \wedge$$

$$\wedge \ \left\{ Z_c t \ ; \ t < v \right\} = V_p[c] \ \wedge$$

$$\wedge \ \sup{}^{v} Z_c = S_p\,[c]$$

$$\wedge \quad \left[\omega^p < v \longrightarrow \left\{Z_c t \; ; \; \omega^p \leq t < v\right\} = V_p\,[d]\right] .$$

Proof: Let $j \geq p$ and $c \in K$ be fixed, let $F_p\,[c] = d$. By lemmata 6.2+3 :

(1) $F_p\,[d] = F_p\,[c]$, $S_p\,[c] = S_p\,[d] = D_c = V_p\,[d] \subsetneq V_p\,[c]$.

Now apply induction on v . Since the formula of the lemma is trivial for $v = 0$, suppose it to be proved for all $u < v$ where $0 \neq v \in Lm_j$.

1.case: $lm_p(v)$. By lemma 6.4 exists $x \in Lm_p$ s.t. $v = x + \omega^p$, $x < v$. Then by induction hypothesis and lemma 6.1 , (c) + (e) + (j):

$$Z_c v = Z_c(x + \omega^p) = F_p^{\,2}\,[c] = F_p\,[c] ,$$

$$\left\{Z_c t \; ; \; t < v\right\} = \left\{Z_c t \; ; \; t < x\right\} \cup \left\{Z_c t \; ; \; x \leq t < x + \omega^p\right\} =$$

$$= V_p\,[c] \cup V_p\,[d] = V_p\,[c] ,$$

$$\sup{}^{v} Z_c = \sup{}^{x + \omega^p} Z_c = S_p\,[d] = S_p\,[c] ,$$

$$\omega^p < v \longrightarrow \left\{Z_c t \; ; \; \omega^p \leq t < v\right\} =$$

$$= \left\{Z_c t \; ; \; \omega^p \leq t < x\right\} \cup \left\{Z_c t \; ; \; x \leq t < x + \omega^p\right\} =$$

$$= \begin{cases} \emptyset \cup D_c & ; \; x = \omega^p \\ V_p\,[d] \cup V_p\,[d] & ; \; x > \omega^p . \end{cases}$$

2.case: $\neg\, lm_p(v)$. By $Lm_j(v)$ we get $Lm_{p+1}(v)$, i.e.

(2) $(\forall y)^{v} \, (\exists z \in Lm_p)^{v} \; y < z$.

For $z < v$, $z \in Lm_p$ we have by induction hypothesis

$$\left\{Z_c t \; ; \; t < z\right\} = V_p\,[c] .$$

Therefore by (2)

$$\left\{Z_c t \; ; \; t < v\right\} = V_p\,[c] .$$

Similarly for $z \in Lm_p$, $\omega^p < z < v$

$$\left\{Z_c t \; ; \; \omega^p \leq t < z\right\} = V_p\,[d] .$$

and thus $\sup^{v} Z_c \subseteq V_p[d]$.
Now let $y < v$. From $\mathrm{Lm}_{p+1}(v)$ follows easily $y + \omega^p < v$. Therefore by lemma 6.1,(e):

$$V_p[d] = \{Z_c t ; y \leq t < y + \omega^p\} \subseteq \{Z_c t ; y \leq t < v\},$$

thus $V_p[d] \subseteq \sup^{v} Z_c$, and thus $\sup^{v} Z_c = V_p[d]$. This yields by lemma 6.2

$$Z_c v = \mathcal{G}[\sup^{v} Z_c] = \mathcal{G}[V_p[d]] = \mathcal{G}[D_c] = F_p[c]. \quad \text{Q.e.d.}$$

Now we are able to prove the completeness of A_α. We keep fixed the sentence ϕ and the notation of lemmata 6.1 - 3. Let $\alpha < \omega^\omega$ be written in the form

$$\alpha = \sum_{i=k}^{o} \omega^i \cdot m_i .$$

By corollary 6.2 we have:

(1) $\quad \Sigma_\alpha\{v\}[w] \wedge Z_c v = d \longrightarrow Z_c w = F_o^{m_o} \dots F_k^{m_k}[d]$,

especially

(2) $\quad \Sigma_\alpha[w] \longrightarrow Z_c w = F_o^{m_o} \dots F_k^{m_k}[c]$.

We want to apply corollary 2.2 to get the absolute form Ψ of a relativized version $\Psi[w]$ of (1) and (2). To this end we have to write the right side of the implications of (1) and (2) in relativized form. (1) and (2) are trivial for $\alpha = 0$. So let $\alpha \neq 0$. Then we can rewrite α as

(3) $\quad \alpha = \sum_{i=k}^{l} \omega^i \cdot n_i + \omega^l$,

where $n_i = m_i$ for $i \neq l$, $n_l = m_l - 1$, and $l \leq k$. From (1) and (2) we get by lemma 6.1, (j):

(4) $\quad \Sigma_\alpha\{v\}[w] \wedge Z_c v = d \longrightarrow \sup^{w} Z_c = S_l F_l^{n_l} \dots F_k^{n_k}[d]$,

(5) $\quad \Sigma_\alpha[w] \longrightarrow \sup^{w} Z_c = S_l F_l^{n_l} \dots F_k^{n_k}[c]$.

Note that

$$(6) \qquad \sup{}^{w} Z_c = e \longleftrightarrow (\sup Z_c = e)\,[w]\,.$$

Now we define for α as in (3)

$$(7) \qquad Q_\alpha[c] =_{df} S_1 F_1^{n_1} \ldots F_k^{n_k}[c]\,.$$

Then we get from (4)-(6) by corollary 2.2 :

<u>Lemma 6.6 :</u> Let $\alpha < \omega^\omega$:

(a) $\qquad \vdash \sum_\alpha \{v\} \wedge Z_c\, v = d \longrightarrow \sup Z_c = Q_\alpha[\,d\,]$,

(b) $\qquad \vdash \sum_\alpha \longrightarrow \sup Z_c = Q_\alpha[\,c\,]$.

The second part of this lemma yields immediately the completeness of $\bar{A}_\alpha$ for $\alpha < \omega^\omega$: Since for $\alpha < \omega^\omega$ trivially $A_\alpha \vdash \Sigma_\alpha$, we get:

$$\text{For } \alpha < \omega^\omega : \; A_\alpha \vdash \sup Z_c = Q_\alpha[\,c\,]\,.$$

Thus by corollary 6.1

$$A_\alpha \vdash \Gamma_c(Z) \longrightarrow \Big[\,\phi \longleftrightarrow \mathcal{L}[Q_\alpha[\,c\,]]\Big]\,.$$

$(\exists Z)\,\Gamma_c(Z)$ is derivable by the recursion lemma 3.8 , and $\mathcal{L}[Q_\alpha[c]]$ is a propositional formula; thus either $\mathcal{L}[Q_\alpha[\,c\,]]$ or $\neg\mathcal{L}[Q_\alpha[c]]$ is derivable. Therefore $A_\alpha \vdash \phi$ or $A \vdash \neg\phi$, what proves the completeness of $\bar{A}_\alpha$ for $\alpha < \omega^\omega$ by theorem 5.1 .

Next we suppose $\alpha = \omega^\omega \cdot \mu$ for some $\mu \neq 0$. Then trivially

$$A_\alpha \vdash IM_p\,,$$

where p is the number associated with ϕ by lemma 6.3 . By applying corollary 2.2 to lemma 6.5 we get

$$\vdash IM_p \longrightarrow \sup Z_c = S_p[\,c\,]\,.$$

Thus we get as above

$$A_\alpha \vdash \Gamma_c(Z) \longrightarrow \Big[\,\phi \longleftrightarrow \mathcal{L}[S_p[\,c\,]]\Big]\,,$$

which proves the completeness of $\bar{A}_\alpha$.

Finally let

(8) $\alpha = \omega^\omega \cdot \mu + \delta$ where $\mu \neq 0$ and $\delta = \sum_{i=k}^{1} \omega^i \cdot m_i + \omega^1 < \omega^\omega$.

<u>Lemma 6.7 :</u> For any $j \geq 0$:

$$A_\alpha \vdash (\exists v \in Lm_j) \ \Sigma_\delta\{v\} .$$

<u>Proof:</u> Assume k choosen as small as possible, i.e. $m_k \neq 0$ if $1 < k$. Then $A_\alpha \vdash T_{k,1}$, i.e.

$$A_\alpha \vdash (\exists x \in Lm_k)\ (\forall y \in Lm_{k+1})\ y < x ,$$

and therefore

(9) $A_\alpha \vdash (\exists x)\ (\forall y \in Lm_{k+1})\ y \leq x$.

Let v be the smallest number s.t.

(10) $(\forall y \in Lm_{k+1})\ y \leq v$.

The existence of v is derivable in A_α by (9). Since $\Sigma^{1}_{m_k,\ldots,m_1}$ belongs to A_α , we get easily from (10) and the definition of $T_{i,q}$

$$A_\alpha \vdash \Sigma^{1}_{m_k,\ldots,m_1}\ \{v\} .$$

Further

$$\vdash LM_i \longleftrightarrow LM_i\ \{x\} ,$$

which yields

$$A_\alpha \vdash LM_1\ \{v\} .$$

Finally

$$A_\alpha \vdash \neg S_{k+1}\ \{v\}$$

follows from (10), which proves

$$A_\alpha \vdash \Sigma_\delta\ \{v\} .$$

Since v is choosen minimal, we get from (10)

$$A_\alpha \vdash (\forall w)^v\ (\exists y \in Lm_{k+1})^v\ w < y ,$$

thus for $j \leq k+2$

(11) $\quad A_\alpha \vdash \mathrm{Lm}_j(v)$.

Now let (11) be proved for any $j \geq k+2$. Assume $\neg \mathrm{Lm}_{j+1}(v)$. Then by (10)

$$(\forall y \in \mathrm{Lm}_{j+1})\ y < v .$$

Since $\mathrm{Lm}_j(v)$, this yields $T_{j,1}$. But $\neg T_{j,1}$ is contained in A_α , which proves (11) for $j+1$. Induction on j yields the lemma. Q.e.d.

Now let p be the number associated with ϕ by lemma 6.3 . We rewrite α as

(12) $\quad \alpha = \omega^p \cdot v + \beta$ where $\nu \neq 0$ and $\beta = \sum_{i=p-1}^{l} \omega^i \cdot n_i + \omega^l$.

If $\beta = 0$, i.e. $l > p$, we have $A_\alpha \vdash \mathrm{LM}_p$, from which follows the completeness of $\bar{A}_\alpha$ as in the preceding case. Thus we can assume $l \leq p$ and thus $\beta \neq 0$.

<u>Lemma 6.8 :</u> $\quad A_\alpha \vdash (\exists v \in \mathrm{Lm}_p)\ \Sigma_\beta\{v\}$

<u>Proof:</u> Let k be the number from (8).

<u>1.case:</u> $k < p$. Then

$$n_i = \begin{cases} m_i \ ; & i = 1,\ldots,k \\ 0 \ ; & i = k+1,\ldots,p-1 \end{cases}$$

Therefore

$$A_\alpha \vdash (\Sigma_\delta\{v\} \longleftrightarrow \Sigma_\beta\{v\}) ,$$

and the lemma follows from lemma 6.7 .

<u>2.case:</u> $k \geq p$. Then $n_i = m_i$, $i = 1,\ldots,p-1$. Thus $\Sigma^{l}_{n_{p-1},\ldots,n_1}$ and LM_l are contained in A_α . Let v be the smallest number s.t.

$$(\forall y \in \mathrm{Lm}_{p-1})\ y \leq v .$$

Then as in the proof of lemma 6.7

$$\mathrm{Lm}_p(v) \wedge \Sigma_\beta\{v\} .$$

Q.e.d.

If we combine lemmata 6.8, 6.6, 6.5 and 6.3 , we get

$$A_\alpha \vdash \sup Z_c = Q_\beta\left[F_p[c]\right],$$

which proves the completeness of $\overline{A}_\alpha$ as in the previous cases. This completes the proof of theorem 2.3 .

Now we want to prove theorem 2.1 . Still we keep fixed the sentence ϕ and the numbers p and h_i, k_i for $i = 0,\dots,p$ from lemmata 6.3 and 6.1 . We define a function

$$\mathrm{red}_\phi : \omega_1 - \{0\} \longrightarrow \omega^\omega - \{0\}$$

as follows: For

$$\alpha = \omega^p \cdot \mu ,\ \mu \neq 0 ,$$

we put $\mathrm{red}_\phi(\alpha) =_{df} \omega^p$. For

$$\alpha = \omega^p \cdot \nu + \sum_{i=p-1}^{l} \omega^i \cdot m_i + \omega^l ,\ l < p ,\ \nu \geq 0 ,$$

we define numbers n_i for $i = 1,\dots,p$ as follows:

$$n_i \begin{cases} = m_i \ ; \ m_i < k_i \\ \equiv m_i \ (k_i - h_i) \wedge h_i \leq n_i < k_i \ ; \ m_i \geq k_i \end{cases} \quad \text{for } i = 1,\dots,p-1 ,$$

$$n_p = \begin{cases} 0 \ ; \ \nu = 0 \\ 1 \ ; \ \nu = 0 . \end{cases}$$

Then we put

$$\mathrm{red}_\phi(\alpha) =_{df} \sum_{i=p}^{l} \omega^i \cdot n_i + \omega^l .$$

From lemmata [Bü] 4.7 + 4.11 follows immediately $Z_c\alpha = Z_c(\mathrm{red}_\phi(\alpha))$, and using lemmata 6.1 and 6.3 we can compress our proof of theorem 2.3 into

$$A_\alpha \vdash \sup Z_c = Q_{\mathrm{red}_\phi(\alpha)}[c] .$$

Therefore:

<u>Theorem 6.1 :</u> $A_\alpha \vdash \phi \longleftrightarrow \mathcal{L}\left[Q_{\mathrm{red}_\phi(\alpha)}[c]\right]$.

We define

$$R(\phi) =_{df} \text{image}(\text{red}_\phi) = \{\text{red}_\phi(\alpha) \; ; \; 0 < \alpha < \omega_1\} .$$

$R(\phi)$ is a finite set of ordinals. The following is an analogue to theorem 6.1 for all countable ordinals:

<u>Theorem 6.2 :</u> $\mathcal{C}_0 \vdash \phi$ iff $\mathcal{C}_0 \vdash \bigwedge_{\alpha \in R(\phi)} \mathcal{L}[Q_\alpha[c]]$.

Theorem 2.1 follows immediately from theorems 6.1 + 2 as follows: Let the sentence ϕ belong to MT [co] , i.e. true in all standard models $[\alpha]$ for $\alpha < \omega_1$. Then by theorems 2.3 and 6.1 , $\mathcal{L}[Q_{\text{red}_\phi(\alpha)}[c]]$ is true for any $\alpha < \omega_1$, i.e. $\mathcal{L}[Q_\alpha[c]]$ is true for any $\alpha \in R(\phi)$. Now $\mathcal{L}[Q_\alpha[c]]$ is a propositional formula. Therefore the right side of theorem 6.2 holds; thus ϕ is derivable from $\mathcal{C}_0$. Note that

$$\mathcal{C}_0 \vdash \phi \longleftrightarrow \bigwedge_{\alpha \in R(\phi)} \mathcal{L}[Q_\alpha[c]] ,$$

which seems to correspond better to theorem 6.1 , is wrong if neither $\mathcal{C}_0 \vdash \phi$ nor $\mathcal{C}_0 \vdash \neg\phi$.

To prove the "only if" -part of theorem 6.2 , assume the negation of the right side. Then as above, $\mathcal{L}[Q_\alpha[c]]$ is false for some $\alpha \in R(\phi)$. By theorem 6.1 , $\mathcal{A}_\alpha \vdash \neg\phi$. Since $\mathcal{A}_\alpha$ is a consistent extension of $\mathcal{C}_0$, $\mathcal{C}_0 \nvdash \phi$. To prove the "if" -part, we need some lemmata.

<u>Lemma 6.9 :</u> Let $l \leq j$:

$\mathcal{C}_0 \vdash \text{lm}_l \longrightarrow (\exists z \in \text{Lm}_j)(\forall y \in \text{Lm}_j) \, y \leq z$.

<u>Proof:</u> Suppose lm_l . By applying corollary 2.2 to lemma 6.4 we get

$$(\exists v \in \text{Lm}_l) \; \Sigma_{\omega^l} \{v\} ,$$

therefore

$$(\exists v)(\forall y \in \text{Lm}_l) \, y \leq v ,$$

and thus

$$(\exists v)\ (\forall y \in Lm_j)\ y \leq v \ .$$

Let z be the smallest number s.t.

$$(\forall y \in Lm_j)\ y \leq z \ ,$$

i.e.

$$(\forall y \in Lm_j)\ y \leq z \ \wedge (\forall v)^{z}\ (\exists y \in Lm_j)\ v < y \ .$$

Assume: $\neg Lm_j\ (z)$. Then

$$(\forall v)^{z}\ (\exists y \in Lm_j)^{z}\ v < y \ ,$$

i.e. $Lm_{j+1}\ (z)$, contradiction. Q.e.d.

Lemma 6.10 : Let $1 < p$:

$$\mathscr{C}_0 \vdash lm_1 \longrightarrow (\exists x_1 \in Lm_1, \ldots, x_p \in Lm_p) \Big\{ \sum\nolimits_{\omega}1 \{x_1\} \ \wedge$$

$$\wedge \bigwedge_{j=1}^{p} \ (\forall y \in Lm_j)\ y \leq x_j \quad \wedge$$

$$\wedge \bigvee_{n_1 < k_1, \ldots, n_{p-1} < k_{p-1}} \ \bigwedge_{j=1}^{p-1} \ (x_j = x_{j+1} + n_j \quad \vee$$

$$\vee \ \Big[x_{j+1} + h_j \leq x_{j+1} + n_j < x_j \quad \wedge$$

$$\wedge \quad x_j \equiv x_{j+1} + n_j\ (k_j - h_i) \Big] \Big) \Big[Lm_j \Big] \Big\} \ .$$

Proof: By relativizing to Lm_j the formula (1) in the proof of lemma 6.1 , we get:

$$v < w \wedge Lm_j\ (v) \wedge Lm_j\ (w) \wedge \neg (\exists y)_{v'}^{w'}\ Lm_{j+1}(y) \longrightarrow$$

$$\longrightarrow \bigvee_{i=0}^{n-1} \ (w \equiv v{+}i\ (n)) \Big[Lm_j \Big] \ .$$

If we combine this formula with lemma 6.9 , we get lemma 6.10 . Q.e.d.

Lemma 6.11 : $\mathscr{C}_0 \vdash \bigvee_{\alpha \in R(\phi)} \sup Z_c = Q_{\alpha}[c]$.

Proof: By induction on p we have

$$Lm_p \vee \bigvee_{j=0}^{p-1} \ lm_j \ .$$

If Lm_p , we get

$$\sup Z_c = S_p [c] = Q_{\omega^p} [c] ,$$

by applying corollary 2.2 to lemma 6.5 . Obviously $\omega^p \in R(\phi)$.
If lm_l , $l < p$, let $x_1, \dots, x_p, n_1, \dots, n_{p-1}$ satisfy lemma 6.10 .
Let

$$n_p =_{df} \begin{cases} 0 ; x_p = o \\ 1 ; x_p \neq o , \end{cases}$$

define $\alpha = \sum_{i=p}^{l} \omega^i \cdot n_i + \omega^l \in R(\phi)$. By relativizing to Lm_i the formula (2) in the proof of lemma 6.1 , we get for $h_i \leq j < k_i$:

$$Lm_i (v) \wedge Z_c v = d \wedge Lm_i (t) \wedge (v + h_i \leq t \wedge$$
$$\wedge \ t = v + j(k_i - h_i)) \left[Lm_i\right] \longrightarrow Z_c t = F_i^j [d] .$$

This yields by lemmata 6.5, 6.6, and 6.10

$$\sup Z_c = S_l F_l^{n_l} \dots F_p^{n_p} [c] = Q_\alpha [c] .$$

Now we prove easily the "if" -part of theorem 6.2 : By lemma 6.11

$$\mathcal{C}_0 \vdash \bigwedge_{\alpha \in R(\phi)} \mathcal{L}\left[Q_\alpha [c]\right] \longrightarrow \mathcal{L}[\sup Z_c]$$

Theorem 6.1 yields

$$\mathcal{C}_0 \vdash \bigwedge_{\alpha \in R(\phi)} \mathcal{L}\left[Q_\alpha [c]\right] \longrightarrow \phi .$$

7. The axiom system $\overline{\mathcal{C}}_1$ for MT $[\omega_1]$

At the end of section 1 of [Bü] we find the complete standard axiom systems Kap_n for ω_n. It is a nice exercise in reading relativized formula, to translate Kap_1 into ordinary notation. If we recall the definition of ACC in section 2, we get

Theorem 7.1*: $\mathcal{C}_1 =_{df} \{WLO, (\forall x)\, ACC[x], \neg ACC\}$ is a complete finite standard axiom system for MT $[\omega_1]$.

The decidability proof in section [Bü] 6 shows at once that $\mathcal{C}_1$ cannot be an axiom system: the decision procedure involves α-splicing for $\alpha < \omega_1$ (via the subset- construction of section [Bü] 4) and for $\alpha = \omega_1$, and further the fact that $\mathcal{P}(\omega_1)/\mathfrak{J}_1$ has no atoms. We will formalize the proof of [Bü] 5+6, thus showing that $\mathcal{C}_1$ becomes an axiom system for MT $[\omega_1]$ if we extend it by the above assumptions. In section 8 we will show that we cannot do without SPLICE, and we will discuss other possible theories MT $[\omega_1]$ in case the axiom of choice fails.

We define

$$Clos\,(u) \equiv_{df} (\forall x)\,(Lm(x)\,[U] \longrightarrow Ux),$$

U is closed, i.e. contains all its limits. The collections $\mathfrak{J}_1$ and σ_1 are defined as in section [Bü] 5:

$$U \in \mathfrak{J}_1 \equiv_{df} (\exists V \subseteq U)\left[Clos\,(V) \wedge Cof\,(V)\right],$$

$$U \in \sigma_1 \equiv_{df} \widetilde{U} \in \mathfrak{J}_1 .$$

The sets $U \notin \sigma_1$ are called stationary. Note that U is stationary iff $U \cap V \neq \emptyset$ for all closed cofinal V. Finally we define

$$At\,(U) \equiv_{df} U \notin \sigma_1 \wedge (\forall V \subseteq U)\left[V \in \sigma_1 \vee U - V \in \sigma_1\right],$$

U is an atom in the Boolean algebra $\mathcal{P}(\omega_1)/\mathfrak{J}_1$.

Theorem 7.1 : $\overline{\mathcal{C}}_1 =_{df} \mathcal{C}_1 \cup \{SPLICE, \neg(\exists U)\, At\,(U)\}$ is a complete axiom system for MT $[\omega_1]$ if we assume the axiom of choice.

For the proof first we have to formalize parts of section [Bü] 5 .

<u>Lemma 7.1 :</u> " $\mathfrak{J}_1$ is a filter" is derivable from $\mathcal{C}_1$.

<u>Proof:</u> We formalize the proof of lemma [Bü] 5.1 for two sets. Thus let X_1, X_2 be cofinal-closed, let $X = X_1 \wedge X_2$. It is trivial to derive that X is closed. To show that X is cofinal, let $w \in X_1$ be given. Define sets Z_1, Z_2 by simultaneous recursion

$$(1) \quad \begin{aligned} Z_1 y &\longleftrightarrow \left[X_1 y \wedge (\exists u \in Z_2)^y \, (\forall v \in Z_1)^y \, v < u\right] \vee y = w , \\ Z_2 y &\longleftrightarrow \left[X_2 y \wedge (\exists u \in Z_1)^y \, (\forall v \in Z_2)^y \, v < u\right] . \end{aligned}$$

It is easy to derive

$$(2) \quad Z_i y \longrightarrow X_i y \wedge y \geq w ,$$

$$(3) \quad i \neq j : Z_i y \wedge Z_i z \wedge y < z \longrightarrow (\exists u \in Z_j) \, y < u < z ,$$

$$(4) \quad i \neq j : Z_i y \longrightarrow \neg Z_j y \wedge (\exists u \in Z_j) \, y < u ,$$

$$(5) \quad Z_i y \longrightarrow (\exists z \in Z_i) \, y < z ,$$

$$(6) \quad \neg (\exists x \in Z_i) \, \mathrm{Lm}\,(x) \, [Z_i] .$$

From (6) we get by $\neg$ACC that $\neg \mathrm{Cof}\,(Z_i)$, $i = 1,2$. Therefore using (4) and (1) there is an $x > w$ s.t . $\mathrm{Cof}\,(Z_i)\,[x]$, $i = 1,2$. By (5) $\mathrm{Lm}\,(x)\,[Z_i]$, $i = 1,2$, and thus by (2) $\mathrm{Lm}\,(x)\,[X_i]$, $i = 1,2$. Since X_i are closed, we get $x \in X_i$, $i = 1,2$; thus $x \in X$. Q.e.d.

<u>Corollary 7.1:</u> $\mathcal{C}_1 \vdash X \in \mathfrak{J}_1 \wedge Y \notin \mathfrak{O}_1 \longrightarrow X \wedge Y \notin \mathfrak{O}_1$.

This is well-known from the theory of Boolean algebras.

Note that the fact that $\mathfrak{J}_1$ is an ω-filter cannot be expressed in the language of MT $[\omega_1]$.

We define <u>the derivative of</u> X ,

$$X' = \{x \, ; \, \mathrm{Lm}\,(x)\,[X]\}$$

<u>Lemma 7.2 :</u> The remarks [Bü] 5.3 + 4 are derivable from $\mathscr{C}_1$.

<u>Proof:</u> As in [Bü] . Similarly to lemma 7.1 we define the set $U = \{x_i \; ; \; i < \omega\}$ by recursion:

$$Uy \longleftrightarrow \left[Xy \wedge (\exists v \in U)^{y} (\forall w \in U)^{y} \, w \leq v \right] \vee y = u \, .$$

Then $x > u$ s.t. $\mathrm{Lm}(x)[U]$ exists by $\neg$ ACC , which derives 5.3 . The derivation of 5.4 is obvious from [Bü] . Q.e.d.

We introduce the $\mathfrak{I}_1$-quantifiers as in [Bü] :

$$(\forall_1 y)\, \phi(y) \equiv_{df} \{ y \; ; \; \phi(y) \} \in \mathfrak{I}_1 \, ,$$

$$(\exists_1 y)\, \phi(y) \equiv_{df} \{ y \; ; \; \phi(y) \} \notin \sigma_1 \, ,$$

$$\mathfrak{X}[\sup_1 Z] \equiv_{df} \bigvee_{\{D \, ; \, \mathfrak{X}[D]\}} \left[(\forall_1 t) \bigvee_{e \in D} Zt = e \wedge \bigwedge_{e \in D} (\exists_1 t)\, Zt = e \right] .$$

We get easily:

<u>Lemma 7.3 :</u> Lemma [Bü] 5.8 is derivable from $\mathscr{C}_1$.

Now we will formalize the decidability proof of section [Bü] 6 . Thus in the following we consider the transition system $\Gamma(X,Z)$ of [Bü] 6.(1) and use the terminology of that section.

<u>Lemma 7.4 :</u> Lemma [Bü] 6.1 is derivable from $\mathscr{C}_1$:

a) $\bigvee_{s \in S} (\forall_1 y)\, S_u y = s$,

b) $(\forall_1 y)\, S_u y = s_1 \wedge (\forall_1 y)\, S_v y = s_2 \longrightarrow$
$[s_1 = s_2 \longleftrightarrow u \approx v \; (-\omega_1)]$.

Note that the uniqueness of the state s in a) follows from b).

<u>Proof:</u> The beginning of the proof of [Bü] shows immediately that $(\forall_1 x)\, S_o x = s_o$ is derivable. For the remark concerning S_u note that

$$(\forall u) \bigvee_{i=0}^{g} (\exists v)\, (\forall y)_v \;\; S_u y = S_i\, y$$

follows from the construction of section [Bü] 4 , see formula [Bü] 4.(8) . Then part a) follows from lemma 7.2 . For b) we use lemma 7.1 .

Q.e.d.

The formulas $\Omega_{s_o,T}$ of [Bü] 6.(3) are not directly expressible in the MT-language. We will use slightly different formulas. So let s_o be a state, $T = \{s_1,\dots,s_n\}$ be a set of states. For any formula ϕ we will write

$$(\exists P_s)_{s \in T} \ \phi \equiv_{df} (\exists P_{s_1},\dots,P_{s_n}) \ \phi ,$$
$$(\exists w_s)_{s \in T} \ \phi \equiv_{df} (\exists w_{s_1},\dots,w_{s_n}) \ \phi .$$

Then we define the formula $\Omega'_{s_o,T}$ as

$$(\exists w_s)_{s \in T} \Big[\bigwedge_{s \in T} (\exists_1 y)\ y \approx w_s \ (-\omega_1) \ \wedge$$
$$\wedge \ (\forall_1 y) \bigvee_{s \in T} y \approx w_s \ (-\omega_1) \wedge (\forall_1 y)\ S_o y = s_o \ \wedge$$
$$\wedge \bigwedge_{s \in T} (\forall_1 y)\ S_{w_s} y = s \Big] .$$

We will see that the $\Omega'_{s_o,T}$ are somewhat easier to handle than the $\Omega_{s_o,T}$.

<u>Lemma 7.5 :</u> Lemma [Bü] 6.2 is derivable from $\mathcal{C}_1$:

$$\bigvee_{s_o,T} \ \Omega'_{s_o,T}$$

<u>Proof:</u> As in [Bü] . Let g be the index of the relation $\approx(-\omega_1)$. Then we derive

$$\bigvee_{k=1}^{g} (\exists w_1,\dots,w_k) \Big[\bigwedge_{i \neq j} w_i \not\approx w_j (-\omega_1) \wedge \bigwedge_{i=1}^{k} (\exists_1 y)\ y \approx w_i (-\omega_1) \ \wedge$$
$$\wedge \ (\forall_1 y) \bigvee_{i=1}^{k} y \approx w_i \ (-\omega_1) \Big] .$$

The lemma follows directly from lemma 7.4 .

Q.e.d.

<u>Lemma 7.6 :</u> The equivalence [Bü] 6.(4) is derivable from $\mathcal{C}_1$:

$$\Omega'_{s_o,T} \longleftrightarrow (\exists P_s)_{s \in T} \Big\{ (\forall y) \bigwedge_{s \in T} \left[P_s y \longrightarrow S_o y = s_o \right] \wedge$$
$$\wedge \ (\forall y)\ (\forall v)^y \bigwedge_{s,r \in T} \left[P_s v \wedge P_r y \longrightarrow S_v y = s \right] \wedge$$
$$\wedge \bigwedge_{s \in T} (\exists_1 y) P_s y \wedge (\forall_1 y) \bigvee_{s \in T} P_s y \Big\} .$$

Proof: Similar to [Bü] . Suppose $\Omega'_{s_o,T}$. From the existence of the $w_s, s \in T$, in $\Omega'_{s_o,T}$ follows the existence of $P_s, s \in T$, and $P =_{df} \bigcup_{s \in T} P_s$ s.t.

(a') $\qquad P_s \notin \mathfrak{O}_1$, $s \in T$,

(b') $\qquad P_s y \longrightarrow y \approx w_s\ (-\omega_1)$, $s \in T$,

(c) $\qquad P \in \mathfrak{I}_1$, $(\forall_1 y)\, S_o y = s_o$, $\bigwedge_{s \in T} (\forall_1 y)\, S_{w_s} y = s$

by comprehension. The properties

(d) $\qquad Py \longrightarrow S_o y = s_o$

(e') $\qquad Py \longrightarrow S_{w_s} y = s$, $s \in T$

follow as in [Bü] (define $\overline{P}$ by COMP , use lemma 7.1). Let $r,s \in T$, $y \in P_s \cap P_r$. By (b') $w_s \approx w_r\ (-\omega_1)$; thus by lemma 7.4(b) and property (c) , $r = s$. I.e.

(f') $\qquad P_s \cap P_r = \emptyset$, $r \neq s$.

As in [Bü] we may assume

(g) $\qquad P$ is cofinal-closed.

Define the set $\overline{P} = \{y_i\}$ by recursion:

$$(1) \qquad \begin{aligned} \overline{P}y \leftrightarrow & \Big\{ Py \wedge (\forall u \in \overline{P})^y\, (\exists t)^y \bigvee_{s \in T} \left[u \approx w_s(t) \wedge P_s u \right] \Big\} \vee \\ & \vee \Big\{ y \neq o \wedge \mathrm{Cof}\, (\overline{P})\, [y] \Big\} . \end{aligned}$$

Then $\overline{P} \in \mathfrak{I}_1$ by (g), (c), (b') . Define $\overline{P}_s = \overline{P} \cap P_s$. By lemma 7.1 and corollary 7.1 follows

(h') $\qquad \overline{P}_s \notin \mathfrak{O}_1$ and $\overline{P}_s y \longrightarrow S_o y = s_o$ for $s \in T$, $\bigcup_{s \in T} \overline{P}_s \in \mathfrak{I}_1$

from (a'), (c), (d) . Let $r,s \in T$, $v < y$ s.t. $\overline{P}_s v \wedge \overline{P}_r y$. Then $\overline{P}_v \wedge \overline{P}_y$.

Thus by (1) there is $q \in T$ and $t < y$ s.t. $v \approx w_q\ (t) \wedge P_q v$. Since $P_s v$, by (f') $q = s$. Therefore $v \approx w_s\ (t)$, and thus $S_v y = S_{w_s} y = s$. This proves

(k') $\qquad v < y \wedge \bar{P}_s v \wedge \bar{P}_r y \longrightarrow S_v y = s$

and thus the right side of the equivalence.

To prove the converse we assume the right side of the equivalence. Then we get P_s and w_s and the statements (m) to (s) as in [Bü]. (m) and (s) imply directly

(u') $\qquad (\forall_1 y)\, S_{w_s} y = s \quad$ for $s \in T$.

Now let $s \in T$ and $P_s v$. By (m) there is $y > v$, $y \in P_s$. By (q) $S_v y = s$. Since $w_s < y$, by (s) $S_{w_s} y = s$. Therefore $S_v y = S_{w_s} y$, i.e. $v \approx w_s\ (-\omega_1)$. This proves

$$P_s v \longrightarrow v \approx w_s\ (-\omega_1) \quad \text{for } s \in T .$$

Now $\Omega'_{s_o,T}$ follows directly from (m), (n), (q). Q.e.d.

<u>Lemma 7.7 :</u> Lemma [Bü] 6.3 is derivable from $\bar{\mathcal{C}}_1$:

$$\Omega_{s_o,T} \;.\!\longrightarrow.\; (\exists Z)\ \Gamma(X,Z) \longleftrightarrow B[s_o,T] .$$

<u>Proof:</u> The proof of [Bü] from left to right translates verbally into a derivation. The set $Y = \{y_i\}$ is defined by recursion. Also the derivation of the other implication follows [Bü] up to the statement that each $v \in P$ belongs to exactly one of the $P_{e,s}$, i.e.

(s') $\qquad Pv \longrightarrow \bigvee_{\langle e,s\rangle \in Q} P_{e,s} v \wedge \bigwedge_{\langle d,r\rangle \neq \langle e,s\rangle} \neg P_{d,r} v .$

Since the e_v, s_v are not expressible, we will work with (s') instead of (s),(t). By (p) we get

(u') $\qquad P_{e,s} v \longrightarrow e \in D_1 .$

Then as in [Bü] we get for every $v,y \in P$, $v < y$, uniquely $e,d \in D_1$ (and $s,r \in T$ s.t. $v \in P_{e,s}$ and $y \in P_{d,r}$) and an $X\,[o,v)$-run $Z[o,v]$ and a D_o-exact $X[v,y)$-run $Z[v,y]$ s.t. $Zo = c$, $Zv = e$,

$Zy = e$. By the uniqueness of e,d these partial runs can be spliced together and yield a total run Z for Γ as in [Bü] . The splicing is accomplished with the splicing axiom as follows: Recall the formula $\phi(y,z,Z)$ which we used in section 5 for the derivation of the subformula [Bü] 4.(4) in order to apply SPLICE . Replace in ϕ the subformula $Zy = d$ by $Zy = e$, replace $\sup^{\bar{x}}$ by $\sup_o^x$ and D by D_o; call the formula $\phi_{e,d}(y,z,Z)$. Then we define the formula $\Psi(v,y,Z)$ as

$$\bigvee_{e,d,s,r} P_{e,s}v \wedge P_{d,r}y \wedge \phi_{e,d}(v,y,Z) .$$

Obviously

$$Y \underset{v}{\overset{y}{\equiv}} Z \longrightarrow \left[\Psi(v,y,Y) \longleftrightarrow \Psi(v,y,Z)\right] .$$

Further, since

$$P_{e,s}v \wedge P_{d,r}y \wedge v < y \longrightarrow (\exists Z)\, \phi_{e,d}(v,y,Z) ,$$

we get from (s')

$$(\forall v,y \in P)\left[(y = v')[P] \longrightarrow (\exists Z)\, \Psi(v,y,Z)\right] .$$

Thus if we apply SPLICE to Ψ, we get a run $Z[z,\omega_1)$ which with COMP can be extended to a total run. Q.e.d.

Note that in the second half of the above proof we have used both, ω_1-splicing and the axiom that $\mathcal{P}(\omega_1)/\mathcal{I}_1$ is atomless. α-splicing for countable α will be used in the next lemma.

<u>Lemma 7.8 :</u> Lemma [Bü] 6.4 is derivable from $\overline{\mathcal{C}}_1$.

<u>Proof:</u> Similar to the derivation of lemma [Bü] 3.5 , see lemma 4.6 . We have to isolate the conditions

$$Pt' \wedge Q_s t' \longrightarrow F^o\left[Xt, \bar{Y}_s t\right] = Y_s^o t' ,$$

$$(\forall_o^x t) \neg Pt \wedge Px \wedge Q_s x \longrightarrow \mathcal{G}^o\left[\sup_o^x \bar{Y}_s\right] = Y_s^o x$$

and to extend the resursion by changing the conditions

$$\neg Q_s t \wedge P_s t \longrightarrow Y_s t = a , \quad P_s t \longrightarrow \bar{Y}_s t = a$$

into

$$\neg Q_s t \longrightarrow Y_s t = \bar{Y}_s t = a \ , \ Pt \longrightarrow \bar{Y}_s t = a \ .$$

Then the argument is as in [Bü] . Q.e.d.

Theorem 7.2 : Theorem [Bü] 6.6 is derivable from $\bar{\mathcal{C}}_1$: To every formula $\Sigma(X)$ of $MT[\omega_1]$ one can construct an automata normal form $(\exists Z)\,\Gamma(X,Z)$ where

$$\Gamma(X,Z) \equiv_{df} E[Zo] \wedge (\forall t)\ H\,[Xt,Zt,Zt'] \wedge (\forall x)\,\mathcal{K}[\sup_o^x Z, Zx] \wedge \wedge\, \mathcal{L}[\sup_1 Z] \ ,$$

s.t. $\bar{\mathcal{C}}_1 \vdash \Sigma(X) \longleftrightarrow (\exists Z)\,\Gamma(X,Z)$.

The proof is immediate from the preceding lemmata. Now we can prove Theorem 7.1 . Let the sentence Σ be in automata normal form $(\exists Z)\,\Gamma(Z)$ where

(1) $\Gamma(Z) \equiv_{df} E[Zo] \wedge (\forall t)\ H\,[Zt,Zt'] \wedge (\forall x)\,\mathcal{K}[\sup_o^x Z, Zx] \wedge \wedge\, \mathcal{L}[\sup_1 Z]$.

We define $S_u v$ as in the beginning of section [Bü] 4 . By lemma [Bü] 4.4 there is a finite-state recursion

(2) $\Gamma'(Y_a) \equiv_{df} Y_a o = a \wedge (\forall t)\ Y_a t' = F\,[Y_a t] \wedge \wedge\, (\forall x)\ Y_a x = \mathcal{B}[\sup_o^x Y_a]$

s.t. for a certain component Y_a^o of Y_a

(3) $Y_a^o v = S_o v$.

Note that by lemma 5.1 the formula (3) is derivable from $\bar{\mathcal{C}}_o$, or more exactly from

WLO , $(\forall x)^{v'}$ ACC [x] , SPLICE ,

i.e. from $\bar{\mathcal{C}}_1$. From (3) we get by corollary 3.2 for every $\alpha < \omega^\omega$

(4) $Y_a^o \alpha = S_o \alpha$.

If we start the resursion Γ' at time v instead of at o , we get $\bar{Y}_a$ s.t. for every $\alpha < \omega^\omega$

(5) $\bar{Y}_a^o \alpha = S_v v + \alpha$.

Formally, if we apply corollary 3.2 to the formula (2) $\longrightarrow$ (4), we get that

$$\Gamma'(\bar{Y}_a)\{v\} \longrightarrow (\bar{Y}_a^o \alpha = S_o \alpha)\{v\}$$

is derivable. This can be restated as

(6) $Y_b v = a \longrightarrow Y_b^o v + \alpha = S_v v + \alpha$.

By the formula (1) in front of lemma 6.6 we have

(7) $Y_b v = a \longrightarrow Y_b v + \alpha = Y_a \alpha$.

Combining (4),(6) and (7), we see that

(8) $S_o \alpha = S_v v + \alpha$

is derivable for every $\alpha < \omega^\omega$. (Note that, similar to lemma 4.7,(8) is all what can be expressed from the trivial equation $S_o t = S_v v + t$.)

Now we introduce for $i < \omega$ the operators

(9) $F_i [c] =_{df} Y_c \omega^i$.

By lemma 6.3 there is a number p s.t.

(10) $F_p^2 = F_p$.

Let

(11) $b =_{df} F_p [a]$, $s =_{df} b^o = Y_a^o \omega^p$.

Let $j \geq p$, let $v > o$ s.t. $Lm_j(v)$. From lemma 6.5 we get

(12) $Y_a v = b$.

Thus by (3) and (11)

(13) $S_o v = s$.

By (12), lemma 6.1(c), and (10)

$$Y_a v + \omega^p = F_p [b] = F_p^2 [a] = F_p [a] = Y_a \omega^p .$$

Thus by (3) and (8)

(14) $\quad S_0\, v+\omega^p = S_0\, \omega^p = S_v\, v+\omega^p$,

which proves $v \approx 0\ (-\omega_1)$. Since the set of limit numbers of order $\geq p$ is obviously cofinal-closed, we get from (13) and (14)

$$(\forall_1 y)\, y \approx o\ (-\omega_1) \wedge (\forall_1 y)\, S_0\, y = s\ .$$

This yields $\Omega'_{s,\{s\}}$. Therefore:

<u>Lemma 7.8 :</u> Remark [Bü] 6.8 is derivable.

From the proof of theorem [Bü] 6.9 follows now that any $MT[\omega_1]$ -sentence is $\bar{\mathcal{C}}_1$ -derivably equivalent to "true" or "false", which completes the proof of theorem 7.1 .

8. The independence of the splicing axiom

$\mathcal{C}_0$ and $\mathcal{C}_1$ are very natural, finite standard axiom systems for MT[co] and MT$[\omega_1]$, resp. We will show now that the splicing axiom schema, which of course holds in MT$[\alpha]$ for all α, is independent from $\mathcal{C}_0$ and $\mathcal{C}_1$. Thus in theorems 2.1 and 7.1, the bar over $\mathcal{C}_0$ and $\mathcal{C}_1$ resp. cannot be removed. The problem of the independence of "$\mathcal{P}(\omega_1)/\mathcal{J}_1$ is atomless" from SPLICE will be discussed afterwards.

We have to show that there are general models of $\mathcal{C}_0$ and $\mathcal{C}_1$ where SPLICE fails. This follows from work of Church [Ch 1] and Hájek [Há]; see also Specker [Sp]. This work can be applied to monadic second order theories by a result of Litman [Li] who shows that MT$[\omega_1] \neq$ MT$[\alpha]$ for $\alpha<\omega_1$ even if ω_1 is ω-accessible.

We follow loosely the terminology of [Ch 1] and [Há]. For a limit ordinal α, the function $f: \omega \to \alpha$ is a fundamental sequence for α iff f is order-preserving and the image of f is cofinal in α, i.e.

$$(\forall n,m<\omega)[n<m \longrightarrow f(n)<f(m)],$$
$$(\forall \beta<\alpha)(\exists n<\omega)\ \beta<f(n).$$

A function $f: \omega \longrightarrow \alpha$ is a fundamental arrangement for α iff f is onto.

Let CS(α) stand for "there is a choice of fundamental sequences on α", i.e. CS(α) holds iff there is a function $f: \alpha \times \omega \to \alpha$ s.t. for all limits $\beta<\alpha$, the function $\lambda n\, f(\beta,n)$ is a fundamental sequence for β. Let CA(α) stand for "there is a choice of fundamental arrangements on α", i.e. CA(α) iff there is a function $f: \alpha \times \omega \to \alpha$ s.t. for all limits $\beta<\alpha$, the function $\lambda n\, f(\beta,n)$ is a fundamental arrangement for β. (Here we use the Church λ-notation for functions to indicate the argument variable.)

Proposition 1 : The axiom of choice AC implies CS(ω_1).

The following propositions can be proved (in ZF set theory) without AC .

<u>Proposition 2 :</u> $CS(\alpha)$ holds for $\alpha < \omega_1$.

<u>Proposition 3 :</u> ([Ch1]) : If $CS(\alpha)$ holds, then $\alpha \leq \omega_1$.

Thus with AC we have: $CS(\alpha)$ holds iff $\alpha \leq \omega_1$.

<u>Proposition 4</u> ([Ch1]) : For all α , $CS(\alpha)$ is equivalent to $CA(\alpha)$.

<u>Proposition 5</u> ([Ch1], theorem A_1 , p. 188) : If ω_1 is ω-accessible, then $CS(\omega_1)$ is false.

Recall that ω_1 is the first uncountable ordinal, i.e. ω_1 is the smallest ordinal α s.t. there is no map $f : \omega \longrightarrow \alpha$ which is onto. By AC , ω_1 is not ω-accessible.

From section 7 or from [Bü] section 5 recall the notion of the derivative U' of a set U :

$$U' = \{x \; ; \; Lm\,(x)[U]\} .$$

Let CD stand for "every closed set of limit numbers is a derivative". CD was introduced in [Li] , and can be stated as an MT-sentence:

$$CD \equiv_{df} (\forall U \subseteq Lm)\left[U \text{ closed} \longrightarrow (\exists V)\; U = V'\right] .$$

<u>Proposition 6</u> ([Li]) : For all $\alpha \leq \omega_1$: $CS(\alpha)$ holds iff $CD \succ\!\!- [\alpha]$.

<u>Proof:</u> For the implication from left to right let $\alpha \leq \omega_1$, let $CS(\alpha)$ be true. Let $f : \alpha \times \omega \longrightarrow \alpha$ be a choice of fundamental sequences on α. Let $U \subseteq \alpha$ be a closed set of limit numbers. Define

$$V =_{df} U \cup \{f(v,n) \; ; \; v \in U \, , \, n < \omega \, , \, ((\exists u < f(v,n)) \; v = u')[U]\} .$$

We want to show: $U = V'$. Thus let $x \in U$.

1.case: $(x = 0 \vee (\exists u)\; x = u')[U]$. Then $x \in V'$, since $\lambda n\, f(x,n)$ is a fundamental sequence for x .

2.case: $\mathrm{Lm}\,(x)[U]$, i.e.

(1) $$(\exists y \in U)\ y < x \wedge (\forall y \in U)^{x} (\exists z \in U)^{x}\ y < z\ .$$

Let $w \in V$, $w < x$. Assume

$$(\forall y \in U)^{x}\ y < w\ .$$

Then by definition of V we are in the 1.case; contradiction. Therefore

$$(\exists y \in U)^{x}\ w \leq y\ .$$

This proves, togehter with (1) and the definition of V

$$(\exists y \in V)\ y < x \wedge (\forall y \in V)^{x} (\exists z \in V)^{x}\ y < z\ ,$$

i.e. $x \in V'$. This shows $U \subseteq V'$.

Now let $x \in V'$, i.e. $\mathrm{Lm}\,(x)[V]$. We have to prove: $x \in U$.

1.case: $(\forall y \in U)^{x} (\exists z \in U)^{x}\ y < z$. Thus either $(x \leq o)[U]$ or $\mathrm{Lm}\,(x)[U]$. If $\mathrm{Lm}\,(x)[U]$, then $x \in U$, since U is closed. Let u_o be the smallest element of U , let $x \leq u_o$. Assume $x < u_o$. Since $\lambda n\, f(u_o,n)$ is a fundamental sequence for u_o , there is a minimal $n < \omega$ s.t. $x \leq f(u_o, n)$. Therefore $f(u_o, j)$, $j = 0,\dots,n-1$, are the only elements of V smaller than x ; thus $\neg \mathrm{Lm}\,(x)[V]$, contradiction. Therefore again $x = u_o \in U$.

2.case: There is $y \in U$, $y < x$ s.t. $(\forall z \in U)^{x}\ z \leq y$. Assume y to be the last element of U . Then $(\forall z \in V)\ z \leq y$; thus $\neg\,\mathrm{Lm}\,(x)[V]$, contradiction. Therefore there is a minimal $z \in U$, $z > y$. Then $z \geq x$. Assume $z > x$. As in 1.case there is a minimal $n < w$ s.t. $x \leq f(z, n)$, and therefore either y or $f(z, n-1)$ is an immediate V-predecessor of x . Therefore again $\neg \mathrm{Lm}\,(x)[V]$, contradiction. This shows $V' \subseteq U$, and thus $CD \rightarrowtail [\alpha]$.

To prove the converse implication, by proposition 2 we have to consider the case $\alpha = \omega_1$ only. Let the pairs of ordinals less than ω_1 be ordered first by maximum and for equal maximum lexicographically.

I.e. for $u,v,y,z < \omega_1$:

$$\langle u,v\rangle \prec \langle y,z\rangle \text{ iff } \max(u,v) < \max(y,z) \vee$$
$$\vee \left[\max(u,v) = \max(y,z) \wedge (u < y \vee [u = y \wedge v < z])\right].$$

This is an ordering of type ω_1 , i.e. there is an order-preserving function $O: \omega_1 \times \omega_1 \longrightarrow \omega_1$ which is (1-1 and) onto. To see that we get O without using AC , let O be defined by recursion: Put $O(o,o) =_{df} o$. If $O(u,v)$ is defined for all $\langle u,v\rangle \prec \langle y,z\rangle$, put

$$O(y,z) =_{df} \min\left\{x \,;\, (\forall \langle u,v\rangle \prec \langle y,z\rangle)\; O(u,v) < x\right\}.$$

We define

$$U =_{df} \left\{O(o,z)\,; z<\omega_1\right\} \cup \left\{O(z,o)\,; z<\omega_1\right\}.$$

Note that for each $z<\omega_1$ the sequence

$$\langle o,z\rangle \prec \langle 1,z\rangle \prec \;\dots\; \prec \langle z,o\rangle$$

enumerates the interval $\left[\langle o,z\rangle , \langle z,o\rangle\right]$, and is of order type $z+1$. Therefore

(1) $U \wedge (O(o,z)\,,\, O(z,o)) = \emptyset$,
$[O(o,z)\,,\, O(z,o))$ is of order type z .

We want to show:

(i) U is a closed set of limits.

(ii) If $U = V'$ for some $V \subseteq \omega_1$, then $CS(\omega_1)$.

(i) and (ii) togehter yields the wanted result.

<u>Proof of (i):</u> Since O is order-preserving, $O(o,z) \geq z$. Therefore U is cofinal in ω_1 . Let $w < \omega_1$ be s.t. $Lm(w)[U]$, i.e.

(2) $$(\forall u \in U)^w (\exists v \in U)^w u < v \wedge (\exists u \in U)\, u < w .$$

Since U is cofinal, there is a smallest $y \in U$ s.t. $y \geq w$. Assume $y = O(z,o)$ for some z . But then by (1) , $O(o,z)$ is an immediate U-predecessor of y , i.e.

$$\neg(\exists v \in U)^y\, O(o,z) < v ,$$

which contradicts (2) . Therefore $y = O(o,x)$; x is a limit by the same argument. By definition of $\prec$ for all u,v

(3) $\langle u,v\rangle \prec \langle o,x\rangle$ iff $u,v < x$.

Now let us assume that $w < y$. By (3), $w = O(u,v)$ for some $u,v < x$. Let $z = \max(u,v)$. Since x is a limit, we get $w < O(o,z+1) < y$. Since $O(o,z+1) \in U$, this is a contradiction to the choice of y. Therefore $w = y$, which proves that U is closed. Since x is a limit, by (3) $\langle o,x\rangle$ is a limit, too. $\langle x,o\rangle$ is a limit by (1). Therefore $U \subseteq Lm$ what proves (i).

<u>Proof of (ii):</u> Assume $U = V'$ for some $V \subseteq \omega_1$. By (1), $V \cap (O(o,z), O(z,o))$ is of order type ω for every z. Thus define a function $g : \omega_1 \times \omega \longrightarrow \omega$ by

$$g(u,n) =_{df} 0 \quad \text{for successor numbers } u \text{ or } u = o,$$
$$g(x,n) =_{df} \min \left\{ z \in V \cap (O(o,x), O(x,o)) \;; (\forall m < n)\, g(x,m) < z \right\}$$

if x is a limit number. Then for all limits x, $\lambda n\, g(x,n) : \omega \longrightarrow V \cap (O(o,x), O(x,o))$ is order-preserving with image cofinal in $O(x,o)$. Let $O_1(u,v) = u$ be the projection on the first component. Define $f = O_1 \circ O^{-1} \circ g$. Then $f : \omega_1 \times \omega \rightarrow \omega_1$. We want to show that f is a choice of fundamental sequences on ω_1. Let $x < \omega_1$ be a limit. Let $n < m < \omega$. Then $g(x,n) < g(x,m) < O(x,o)$. By (1), there are $u,v < x$ s.t. $g(x,n) = O(u,x)$, $g(x,m) = O(v,x)$. Thus $f(x,n) = u < v = f(x,m) < x$. Thus $\lambda n\, f(x,n) : \omega \longrightarrow x$ is order-preserving. Let $y < x$. Then $O(y,x) \in U \cap [O(o,x), O(x,o))$. There is $n < \omega$ s.t. $g(x,n) \geq O(y,x)$. Then $f(x,n) \geq y$, thus the image of $\lambda n\, f(x,n)$ is cofinal in x. Therefore $CS(\omega_1)$ holds, which proves (ii), and thus the proposition.

Q.e.d.

<u>Corollary 1</u> ([Li]): $MT[\omega_1] \neq MT[\alpha]$ for all $\alpha < \omega_1$, even if ω_1 is ω-accessible. In fact, for all $\alpha \leq \omega_1$: $ACC \wedge CD \succ\!\!-\, [\alpha]$ iff $\alpha < \omega_1$.

<u>Proof:</u> If ω_1 is not ω-accessible, then $ACC \notin MT[\omega_1]$.
If ω_1 is ω-accessible, then by proposition 5 $CS(\omega_1)$ fails. There-

fore by propositions 6 and 2, CD$\succ$— $[\alpha]$ iff $\alpha < \omega_1$. Q.e.d.

Recall that ω_1 is <u>singular</u> iff it is the union of countably many countable sets; otherwise it is <u>regular</u>. The following facts are well-known:

<u>Proposition 7 :</u> AC implies that ω_1 is regular. ω_1 is singular iff ω_1 is ω-accessible.

<u>Proposition 8</u> (Levi [Le], for a proof see e.g. Hájek [Há]): "ω_1 is singular" is consistent with ZF.

<u>Corollary 2 :</u> CS(ω_1) is independent from ZF.

By propositions 6 and 5 we get:

<u>Corollary 3</u> ([Li]) : "CD $\in$ MT[ω_1]" is independent from ZF.

<u>Proposition 9 :</u> $\{$WLO, $(\forall x)$ ACC$[x]$, SPLICE$\}$ $\vdash_{\mathcal{P}_2}$ CD, where $\mathcal{P}_2$ is the system of section 1 (which contains no set theory besides Extensionality and Comprehension).

<u>Proof:</u> Let U be a closed set of limit numbers. Define

$$\phi(y,z,Z) \equiv_{df} \mathrm{Lm}(z)[Z] \wedge \neg(\exists u)_y^z\, \mathrm{Lm}(u)[Z],$$

i.e. $\phi(y,z,Z)$ iff Z is an ω-sequence in the intervall $[y,z)$ with limit z. Obviously ϕ satisfies the condition (1) of SPLICE

$$Z \overset{z}{\underset{y}{\equiv}} Y \longrightarrow \left[\phi(y,z,Z) \longleftrightarrow \phi(y,z,Y)\right].$$

From $(\forall x)$ ACC $[x]$ and COMP we get

$$(\forall y,z \in U)\left[(z = y')\,[U] \longrightarrow (\exists Z)\ \phi(y,z,Z)\right].$$

Thus by SPLICE there is a set V s.t.

$$(\forall y,z \in U)\left[(z = y')\,[U] \longrightarrow \phi(y,z,V)\right].$$

It is easy to check that V' = U. Q.e.d.

<u>Corollary 4 :</u> SPLICE is independent from $\mathcal{C}_0$ and from A_α where

$\omega^\omega < \alpha < \omega_1$.

<u>Proof:</u> SPLICE is of course consistent with $\mathcal{C}_0$. If SPLICE were derivable from $\mathcal{C}_0$, by proposition 9 CD would be derivable from $\mathcal{C}_0$. But this contradicts corollary 3 (and its proof). Obviously Levi's ω_1 of proposition 8 is a model of A_{ω^ω}. This shows that SPLICE is independent from A_{ω^ω}. More general, for this ω_1 and any countable ordinal μ, $\omega_1 + \mu$ is a model of $A_{\omega^\omega+\mu}$ which violates CD . Therefore for $\omega^\omega < \alpha < \omega_1$, SPLICE is independent from A_α.Q.e.d.

The independence of SPLICE from $\mathcal{C}_1$ follows from the paper of Hájek [Há] , where the consistency of Church's alternative B is shown:

<u>Proposition 10</u> ([Há] , theorem 5): "ω_1 is regular and $CS(\omega_1)$ fails" is consistent with ZF .

From proposition 10 together with proposition 9 and corollary 3 we get

<u>Corollary 5 :</u> SPLICE is independent from $\mathcal{C}_1$.

From corollaries 4+5 we see easily that of the theories $cl(\mathcal{C}_0)$, $cl(\mathcal{C}_1)$, $cl(A_\alpha)$ where $\omega^\omega < \alpha < \omega_1$, none satisfies the axiom of definable choice. Therefore very likely none of MT[co] and MT[α] where $\omega^\omega < \alpha \leq \omega_1$ satisfy the axiom of definable choice either.

Büchi's decidability proofs for MT[co] and MT[ω_1] use the axiom of choice in the form of the splicing principle to get the automata normal forms. How do we get nice, i.e. decidable, normal forms without AC ? The situation here is quite similar e.g. to infinite cardinal arithmetic or to infinite distributive laws in Boolean algebras. No nice language without the axiom of choice (where a language is a set of expressions describing something). Thus

Problem 1 : If ω_1 is singular, is $MT[\omega_1]$ decible ?

This is not a precise question, since $MT[\omega_1]$ need not be uniquely determined by the statement "ω_1 is singular" (added to ZF). More precisely: If $\mathcal{M}$ is a model of any set theory, let $\omega_1^{\mathcal{M}}$ be the first uncountable ordinal in $\mathcal{M}$. There might be models $\mathcal{M}, \mathcal{N}$ of ZF + "ω_1 is singular" s.t. $MT\left[\omega_1^{\mathcal{M}}\right] \neq MT\left[\omega_1^{\mathcal{N}}\right]$.

Before we discuss this problem further, let us have a look at the second axiom in $\mathcal{C}_1$ which follows from AC : $\neg(\exists U)\, At(U)$, i.e. $\mathcal{P}(\omega_1)/\mathfrak{I}_1$ is atomless. The proof in section [Bü]5 shows:

Proposition 11 : $\mathfrak{I}_1$ is an ω-filter and $CA(\omega_1)$ imply that $\mathcal{P}(\omega_1)/\mathfrak{I}_1$ is atomless.

Lemma [Bü]5.1 is proved without AC . But note that we need AC to derive from it that $\mathfrak{I}_1$ is ω-complete: If $Y_i \in \mathfrak{I}_1$, $i<\omega$, we have to choose for any $i<\omega$ a fixed closed-unbounded $X_i \subseteq Y_i$ to apply lemma [Bü]5.1 . Actually, however, the above quoted proof yields: any stationary set contains ω_1 many disjoint stationary subsets. If the proof could be changed in such a way that "ω-filter" and "ω_1 many" can be replaced by "filter" and "finitely many" resp., then we might be able to translate it into an MT-derivation. Lemma 7.1 and propositions 4 , 6, and 9 therefore suggest to derive $\neg(\exists U)\, At(U)$ from SPLICE .

Problem 2 : Are the following two statements true ?

(a) "CD $\succ\!\!-\ \omega_1$" implies that $\mathcal{P}(\omega_1)/\mathfrak{I}_1$ is atomless.

(b) $\mathcal{C}_1 \cup \{SPLICE\} \vdash \neg(\exists U)\, At(U)$.

By proposition 9 , (a) implies (b) . By propositions 11, 4, and 6, (a) would be true if any general model

$$\mathfrak{D} = \langle D, \mathcal{P}_0(D)\ ; \in\, ,\, =\, ,\, < \rangle$$

of $\mathcal{C}_1 \cup \{CD\}$ could be extended to a model $\mathcal{M}$ of ZF s.t. $\omega_1^{\mathcal{M}} = D$, $\mathcal{P}^{\mathcal{M}}(\omega_1^{\mathcal{M}}) = \mathcal{P}_0(D)$ (thus $CA(\omega_1^{\mathcal{M}})$) and $\mathcal{J}_1^{\mathcal{M}}$ is ω-complete. This, however, seems very unlikely. (Note that proposition 6 is proved in ZF , i.e. holds for ZF-models only.) Whether conversely (b) implies (a) depends on

Problem 3 : $\mathcal{C}_1 \cup \{CD\} \vdash$ SPLICE ?

A positive answer to problem 3 would yield finite axiom systems for MT[co] and MT$[\omega_1]$.

To sum up, there are the following possibilities for MT$[\omega_1]$:

1.case: AC holds. Then ω_1 is regular and CS (ω_1). MT$[\omega_1]$ = cl$(\overline{\mathcal{C}}_1)$ is decidable.

2.case: ω_1 is singular. Then AC fails, ω_1 is ω-accessible, thus $\neg CS(\omega_1)$. Therefore SPLICE does not belong to MT$[\omega_1]$. In Litman [Li] various possibilities are discussed to get $\mathcal{C}_1$ complete (and thus decidable) by statements concerning the atoms of $\mathcal{P}(\omega_1)/\mathcal{J}_1$. It seems that the consistency remains open.

3.case: ω_1 is regular, but AC fails. If $\mathcal{P}(\omega_1)/\mathcal{J}_1$ is atomless and SPLICE holds for ω_1 , then MT$[\omega_1]$ = cl$(\overline{\mathcal{C}}_1)$ is decidable as in case 1 . Otherwise the situation is similar to case 2 . We quote one result: By Jech [Je] , ω_1 can be measurable. Let $\mathcal{M}$ be Jech's ZF-Model. The non-principal ω-complete ultrafilter on $\mathcal{P}(\omega_1^{\mathcal{M}})$ which shows the measurability, contains all closed-cofinal sets. Therefore $\mathcal{P}(\omega_1^{\mathcal{M}})/\mathcal{J}_1$ is the 2-element Boolean algebra, which is atomic. By proposition 11 $\neg CA(\omega_1^{\mathcal{M}})$, thus SPLICE fails in $\omega_1^{\mathcal{M}}$. Note that also the axiom of determinateness implies that $\mathcal{J}_1$ is an ultra-filter (see Mycielski [My]).

With MT[co] the situation is not much better. Remark 2.1 and corollary 4.2 seem to imply that we can show without AC that SPLICE belongs to MT[co] ; i.e. that MT[co] is not changed if we reject AC . The proof of remark 2.1 , however, uses the decision procedure of [Bü] and thus AC . It might be that remark 2.1 becomes false if we do not have the automata normal forms which we get by slicing. The only theories which are surely not affected by AC , are $MT[\alpha] = cl(A_\alpha)$ for $\alpha < \omega^\omega$.

We restate the discussion in the following problems:

<u>Problem 4 :</u> Are the theories (a) $cl(\mathcal{C}_0)$ and (b) $cl(\mathcal{C}_1)$ decidable ?

<u>Problem 5 :</u> What are the consistent and complete (and decidable) extensions of (a) $\mathcal{C}_0$ and (b) $\mathcal{C}_1$ which violate SPLICE ?

<u>Problem 6 :</u> If ω_1 is measurable, is $MT[\omega_1]$ (a) uniquely determined, (b) decidable ?

References

[Bi] Garrett Birkhoff, "Lattice Theory". AMS Coll. Publ. 25, Providence 1963

[Bü 1] J. Richard Büchi, "Weak second order arithmetic and finite automata". Z. Math. Logik Grundl. Math. 6 (1960), 66-92

[Bü 2] J. Richard Büchi, "On a decision method in restricted second order arithmetic". In "Logic Meth. Phil. Sc., Proc. 1960 Stanford Intern. Congr.", Stanford 1962, 1-11

[Bü 3] J Richard Büchi, "Transfinite automata recursions and weak second order theory of ordinals". In "Logic Meth. Phil. Sc., Proc. 1964 Jerusalem Intern. Congr.", Amsterdam 1965, 3-23

[Bü 4] J. Richard Büchi, "Decision methods in the theory of ordinals". Bull. AMS 71 (1965), 767-770

[Bü] J. Richard Büchi, "The monadic second order theory of ω_1". This volume.

[BS 1] J. Richard Büchi, Dirk Siefkes, "The complete extensions of the monadic second order theory of countable ordinals". To appear

[Ch 1] Alonzo Church, "Alternatives to Zermelo's assumption". Trans. AMS 29 (1927), 178-208

[Ch 2] Alonzo Church, "Introduction to mathematical logic. I". Princeton 1956

[Há] Peter Hájek, "The consistency of Church's alternatives". Bull. Acad. Pol. Sc. 14 (1966), 423-430

[He 1] Leo Henkin, "Completeness in the theory of types". J. Symb. Logic 15 (1950), 81-91

[He 2] Leo Henkin, "Banishing the rule of substitution for functional variables". J. Symb. Logic 18 (1953), 201-208

[Je] Thomas J. Jech, "ω_1 can be measurable". Israel J. Math. 6 (1968), 363-367

[Le] Azriel Levi, "I present here an outline ...". Unpublished manuscript

[Li] Ami Litman, "On the decidability of the monadic theory of ω_1". M. Sc. thesis, The Hebrew University, Jerusalem 1972 (Hebrew)

[MT 1] Andrzej Mostowski, Alfred Tarski, "Arithmetical classes and types of well-ordered systems". Bull. AMS 55 (1949), 65 (abstract)

[MT 2] Andrzej Mostowski, Alfred Tarski, "The elementary theory of well ordering". Unpublished.

[My] Jan Mycielski, "On the axiom of determinateness". Fund. Math. 53 (1963/64), 205-224

[Sh] Saharon Shelah, "The monadic theory of order". Institute of Mathematics, The Hebrew University, Jerusalem, Israel, Winter 1972/73

[Si 1] Dirk Siefkes, "Decidable Theories I. Büchi's monadic second order successor arithmetic". Lect. Notes Math. 120, Berlin Heidelberg New York 1970

[Si 2] Dirk Siefkes, "Recursive models for certain monadic second order fragments of arithmetic". To appear

[Sk 1] Thoralf Skolem, "Einige Bemerkungen zur axiomatischen Begründung der Mengenlehre". Proc. 5th Scand. Math. Congr. Helsinki 1922, 217-232

[Sk 2] Thoralf Skolem, "Über einige Grundlagenfragen der Mathematik". Skrifter Vitenskapsakademiet i Oslo I, No. 4 (1929), 1-49

[Sk 3] Thoralf Skolem, "Über die Nichtcharakterisierbarkeit der Zahlenreihe mittels endlich oder abzählbar unendlich vieler Aussagen mit ausschließlich Zahlenvariablen". Fund. Math. 23 (1934), 150-161

[Sp] Ernst Specker, "Zur Axiomatik der Mengenlehre (Fundierungs- und Auswahlaxiom)". Z. Math. Logik Grundl. Math. 3 (1957), 173-210

Lecture Notes in Mathematics

Vol. 146: A. B. Altman and S. Kleiman, Introduction to Grothendieck Duality Theory. II, 192 pages. 1970. DM 18,–

Vol. 147: D. E. Dobbs, Cech Cohomological Dimensions for Commutative Rings. VI, 176 pages. 1970. DM 16,–

Vol. 148: R. Azencott, Espaces de Poisson des Groupes Localement Compacts. IX, 141 pages. 1970. DM 16,–

Vol. 149: R. G. Swan and E. G. Evans, K-Theory of Finite Groups and Orders. IV, 237 pages. 1970. DM 20,–

Vol. 150: Heyer, Dualität lokalkompakter Gruppen. XIII, 372 Seiten. 1970. DM 20,–

Vol. 151: M. Demazure et A. Grothendieck, Schémas en Groupes I. (SGA 3). XV, 562 pages. 1970. DM 24,–

Vol. 152: M. Demazure et A. Grothendieck, Schémas en Groupes II. (SGA 3). IX, 654 pages. 1970. DM 24,–

Vol. 153: M. Demazure et A. Grothendieck, Schémas en Groupes III. (SGA 3). VIII, 529 pages. 1970. DM 24,–

Vol. 154: A. Lascoux et M. Berger, Variétés Kähleriennes Compactes. VII, 83 pages. 1970. DM 16,–

Vol. 155: Several Complex Variables I, Maryland 1970. Edited by J. Horváth. IV, 214 pages. 1970. DM 18,–

Vol. 156: R. Hartshorne, Ample Subvarieties of Algebraic Varieties. XIV, 256 pages. 1970. DM 20,–

Vol. 157: T. tom Dieck, K. H. Kamps und D. Puppe, Homotopietheorie. VI, 265 Seiten. 1970. DM 20,–

Vol. 158: T. G. Ostrom, Finite Translation Planes. IV. 112 pages. 1970. DM 16,–

Vol. 159: R. Ansorge und R. Hass. Konvergenz von Differenzenverfahren für lineare und nichtlineare Anfangswertaufgaben. VIII, 145 Seiten. 1970. DM 16,–

Vol. 160: L. Sucheston, Constributions to Ergodic Theory and Probability. VII, 277 pages. 1970. DM 20,–

Vol. 161: J. Stasheff, H-Spaces from a Homotopy Point of View. VI, 95 pages. 1970. DM 16,–

Vol. 162: Harish-Chandra and van Dijk, Harmonic Analysis on Reductive p-adic Groups. IV, 125 pages. 1970. DM 16,–

Vol. 163: P. Deligne, Equations Différentielles à Points Singuliers Reguliers. III, 133 pages. 1970. DM 16,–

Vol. 164: J. P. Ferrier, Seminaire sur les Algebres Complètes. II, 69 pages. 1970. DM 16,–

Vol. 165: J. M. Cohen, Stable Homotopy. V, 194 pages. 1970. DM 16,–

Vol. 166: A. J. Silberger, PGL_2 over the p-adics: its Representations, Spherical Functions, and Fourier Analysis. VII, 202 pages. 1970. DM 18,–

Vol. 167: Lavrentiev, Romanov and Vasiliev, Multidimensional Inverse Problems for Differential Equations. V, 59 pages. 1970. DM 16,–

Vol. 168: F. P. Peterson, The Steenrod Algebra and its Applications: A conference to Celebrate N. E. Steenrod's Sixtieth Birthday. VII, 317 pages. 1970. DM 22,–

Vol. 169: M. Raynaud, Anneaux Locaux Henséliens. V, 129 pages. 1970. DM 16,–

Vol. 170: Lectures in Modern Analysis and Applications III. Edited by C. T. Taam. VI, 213 pages. 1970. DM 18,–

Vol. 171: Set-Valued Mappings, Selections and Topological Properties of 2^X. Edited by W. M. Fleischman. X, 110 pages. 1970. DM 16,–

Vol. 172: Y.-T. Siu and G. Trautmann, Gap-Sheaves and Extension of Coherent Analytic Subsheaves. V, 172 pages. 1971. DM 16,–

Vol. 173: J. N. Mordeson and B. Vinograde, Structure of Arbitrary Purely Inseparable Extension Fields. IV, 138 pages. 1970. DM 16,–

Vol. 174: B. Iversen, Linear Determinants with Applications to the Picard Scheme of a Family of Algebraic Curves. VI, 69 pages. 1970. DM 16,–

Vol. 175: M. Brelot, On Topologies and Boundaries in Potential Theory. VI, 176 pages. 1971. DM 18,–

Vol. 176: H. Popp, Fundamentalgruppen algebraischer Mannigfaltigkeiten. IV, 154 Seiten. 1970. DM 16,–

Vol. 177: J. Lambek, Torsion Theories, Additive Semantics and Rings of Quotients. VI, 94 pages. 1971. DM 16,–

Vol. 178: Th. Bröcker und T. tom Dieck, Kobordismentheorie. XVI, 191 Seiten. 1970. DM 18,–

Vol. 179: Seminaire Bourbaki – vol. 1968/69. Exposés 347-363. IV. 295 pages. 1971. DM 22,–

Vol. 180: Séminaire Bourbaki – vol. 1969/70. Exposés 364-381. IV, 310 pages. 1971. DM 22,–

Vol. 181: F. DeMeyer and E. Ingraham, Separable Algebras over Commutative Rings. V, 157 pages. 1971. DM 16,–

Vol. 182: L. D. Baumert. Cyclic Difference Sets. VI, 166 pages. 1971. DM 16,–

Vol. 183: Analytic Theory of Differential Equations. Edited by P. F. Hsieh and A. W. J. Stoddart. VI, 225 pages. 1971. DM 20,–

Vol. 184: Symposium on Several Complex Variables, Park City, Utah, 1970. Edited by R. M. Brooks. V, 234 pages. 1971. DM 20,–

Vol. 185: Several Complex Variables II, Maryland 1970. Edited by J. Horváth. III, 287 pages. 1971. DM 24,–

Vol. 186: Recent Trends in Graph Theory. Edited by M. Capobianco/J. B. Frechen/M. Krolik. VI, 219 pages. 1971. DM 18,–

Vol. 187: H. S. Shapiro, Topics in Approximation Theory. VIII, 275 pages. 1971. DM 22,–

Vol. 188: Symposium on Semantics of Algorithmic Languages. Edited by E. Engeler. VI, 372 pages. 1971. DM 26,–

Vol. 189: A. Weil, Dirichlet Series and Automorphic Forms. V. 164 pages. 1971. DM 16,–

Vol. 190: Martingales. A Report on a Meeting at Oberwolfach, May 17-23, 1970. Edited by H. Dinges. V, 75 pages. 1971. DM 16,–

Vol. 191: Séminaire de Probabilités V. Edited by P. A. Meyer. IV, 372 pages. 1971. DM 26,–

Vol. 192: Proceedings of Liverpool Singularities – Symposium I. Edited by C. T. C. Wall. V, 319 pages. 1971. DM 24,–

Vol. 193: Symposium on the Theory of Numerical Analysis. Edited by J. Ll. Morris. VI, 152 pages. 1971. DM 16,–

Vol. 194: M. Berger, P. Gauduchon et E. Mazet. Le Spectre d'une Variété Riemannienne. VII, 251 pages. 1971. DM 22,–

Vol. 195: Reports of the Midwest Category Seminar V. Edited by J.W. Gray and S. Mac Lane. III, 255 pages. 1971. DM 22,–

Vol. 196: H-spaces – Neuchâtel (Suisse)- Août 1970. Edited by F. Sigrist, V, 156 pages. 1971. DM 16,–

Vol. 197: Manifolds – Amsterdam 1970. Edited by N. H. Kuiper. V. 231 pages. 1971. DM 20,–

Vol. 198: M. Hervé, Analytic and Plurisubharmonic Functions in Finite and Infinite Dimensional Spaces. VI, 90 pages. 1971. DM 16,–

Vol. 199: Ch. J. Mozzochi, On the Pointwise Convergence of Fourier Series. VII, 87 pages. 1971. DM 16,–

Vol. 200: U. Neri, Singular Integrals. VII, 272 pages. 1971. DM 22,–

Vol. 201: J. H. van Lint, Coding Theory. VII, 136 pages. 1971. DM 16,–

Vol. 202: J. Benedetto, Harmonic Analysis on Totally Disconnected Sets. VIII, 261 pages. 1971. DM 22,–

Vol. 203: D. Knutson, Algebraic Spaces. VI, 261 pages. 1971. DM 22,–

Vol. 204: A. Zygmund, Intégrales Singulières. IV, 53 pages. 1971. DM 16,–

Vol. 205: Séminaire Pierre Lelong (Analyse) Année 1970. VI, 243 pages. 1971. DM 20,–

Vol. 206: Symposium on Differential Equations and Dynamical Systems. Edited by D. Chillingworth. XI, 173 pages. 1971. DM 16,–

Vol. 207: L. Bernstein, The Jacobi-Perron Algorithm – Its Theory and Application. IV, 161 pages. 1971. DM 16,–

Vol. 208: A. Grothendieck and J. P. Murre, The Tame Fundamental Group of a Formal Neighbourhood of a Divisor with Normal Crossings on a Scheme. VIII, 133 pages. 1971. DM 16,–

Vol. 209: Proceedings of Liverpool Singularities Symposium II. Edited by C. T. C. Wall. V, 280 pages. 1971. DM 22,–

Vol. 210: M. Eichler, Projective Varieties and Modular Forms. III, 118 pages. 1971. DM 16,–

Vol. 211: Théorie des Matroïdes. Edité par C. P. Bruter. III, 108 pages. 1971. DM 16,–

Please turn over

Vol. 212: B. Scarpellini, Proof Theory and Intuitionistic Systems. VII, 291 pages. 1971. DM 24,–

Vol. 213: H. Hogbe-Nlend, Théorie des Bornologies et Applications. V, 168 pages. 1971. DM 18,–

Vol. 214: M. Smorodinsky, Ergodic Theory, Entropy. V, 64 pages. 1971. DM 16,–

Vol. 215: P. Antonelli, D. Burghelea and P. J. Kahn, The Concordance-Homotopy Groups of Geometric Automorphism Groups. X, 140 pages. 1971. DM 16,–

Vol. 216: H. Maaß, Siegel's Modular Forms and Dirichlet Series. VII, 328 pages. 1971. DM 20,–

Vol. 217: T. J. Jech, Lectures in Set Theory with Particular Emphasis on the Method of Forcing. V, 137 pages. 1971. DM 16,–

Vol. 218: C. P. Schnorr, Zufälligkeit und Wahrscheinlichkeit. IV, 212 Seiten 1971. DM 20,–

Vol. 219: N. L. Alling and N. Greenleaf, Foundations of the Theory of Klein Surfaces. IX, 117 pages. 1971. DM 16,–

Vol. 220: W. A. Coppel, Disconjugacy. V, 148 pages. 1971. DM 16,–

Vol. 221: P. Gabriel und F. Ulmer, Lokal präsentierbare Kategorien. V, 200 Seiten. 1971. DM 18,–

Vol. 222: C. Meghea, Compactification des Espaces Harmoniques. III, 108 pages. 1971. DM 16,–

Vol. 223: U. Felgner, Models of ZF-Set Theory. VI, 173 pages. 1971. DM 16,–

Vol. 224: Revètements Etales et Groupe Fondamental. (SGA 1). Dirigé par A. Grothendieck XXII, 447 pages. 1971. DM 30,–

Vol. 225: Théorie des Intersections et Théorème de Riemann-Roch. (SGA 6). Dirigé par P. Berthelot, A. Grothendieck et L. Illusie. XII, 700 pages. 1971. DM 40,–

Vol. 226: Seminar on Potential Theory, II. Edited by H. Bauer. IV, 170 pages. 1971. DM 18,–

Vol. 227: H. L. Montgomery, Topics in Multiplicative Number Theory. IX, 178 pages. 1971. DM 18,–

Vol. 228: Conference on Applications of Numerical Analysis. Edited by J. Ll. Morris. X, 358 pages. 1971. DM 26,–

Vol. 229: J. Väisälä, Lectures on n-Dimensional Quasiconformal Mappings. XIV, 144 pages. 1971. DM 16,–

Vol. 230: L. Waelbroeck, Topological Vector Spaces and Algebras. VII, 158 pages. 1971. DM 16,–

Vol. 231: H. Reiter, L^1-Algebras and Segal Algebras. XI, 113 pages. 1971. DM 16,–

Vol. 232: T. H. Ganelius, Tauberian Remainder Theorems. VI, 75 pages. 1971. DM 16,–

Vol. 233: C. P. Tsokos and W. J. Padgett. Random Integral Equations with Applications to Stochastic Systems. VII, 174 pages. 1971. DM 18,–

Vol. 234: A. Andreotti and W. Stoll. Analytic and Algebraic Dependence of Meromorphic Functions. III, 390 pages. 1971. DM 26,–

Vol. 235: Global Differentiable Dynamics. Edited by O. Hájek, A. J. Lohwater, and R. McCann. X, 140 pages. 1971. DM 16,–

Vol. 236: M. Barr, P. A. Grillet, and D. H. van Osdol. Exact Categories and Categories of Sheaves. VII, 239 pages. 1971. DM 20,–

Vol. 237: B. Stenström. Rings and Modules of Quotients. VII, 136 pages. 1971. DM 16,–

Vol. 238: Der kanonische Modul eines Cohen-Macaulay-Rings. Herausgegeben von Jürgen Herzog und Ernst Kunz. VI, 103 Seiten. 1971. DM 16,–

Vol. 239: L. Illusie, Complexe Cotangent et Déformations I. XV, 355 pages. 1971. DM 26,–

Vol. 240: A. Kerber, Representations of Permutation Groups I. VII, 192 pages. 1971. DM 18,–

Vol. 241: S. Kaneyuki, Homogeneous Bounded Domains and Siegel Domains. V, 89 pages. 1971. DM 16,–

Vol. 242: R. R. Coifman et G. Weiss, Analyse Harmonique Non-Commutative sur Certains Espaces. V, 160 pages. 1971. DM 16,–

Vol. 243: Japan-United States Seminar on Ordinary Differential and Functional Equations. Edited by M. Urabe. VIII, 332 pages. 1971. DM 26,–

Vol. 244: Séminaire Bourbaki – vol. 1970/71. Exposés 382–399. IV, 356 pages. 1971. DM 26,–

Vol. 245: D. E. Cohen, Groups of Cohomological Dimension One. V, 99 pages. 1972. DM 16,–

Vol. 246: Lectures on Rings and Modules. Tulane University Ring and Operator Theory Year, 1970–1971. Volume I. X, 661 pages. 1972. DM 40,–

Vol. 247: Lectures on Operator Algebras. Tulane University Ring and Operator Theory Year, 1970–1971. Volume II. XI, 786 pages. 1972. DM 40,–

Vol. 248: Lectures on the Applications of Sheaves to Ring Theory. Tulane University Ring and Operator Theory Year, 1970–1971. Volume III. VIII, 315 pages. 1971. DM 26,–

Vol. 249: Symposium on Algebraic Topology. Edited by P. J. Hilton. VII, 111 pages. 1971. DM 16,–

Vol. 250: B. Jónsson, Topics in Universal Algebra. VI, 220 pages. 1972. DM 20,–

Vol. 251: The Theory of Arithmetic Functions. Edited by A. A. Gioia and D. L. Goldsmith VI, 287 pages. 1972. DM 24,–

Vol. 252: D. A. Stone, Stratified Polyhedra. IX, 193 pages. 1972. DM 18,–

Vol. 253: V. Komkov, Optimal Control Theory for the Damping of Vibrations of Simple Elastic Systems. V, 240 pages. 1972. DM 20,–

Vol. 254: C. U. Jensen, Les Foncteurs Dérivés de $\varprojlim$ et leurs Applications en Théorie des Modules. V, 103 pages. 1972. DM 16,–

Vol. 255: Conference in Mathematical Logic – London '70. Edited by W. Hodges. VIII, 351 pages. 1972. DM 26,–

Vol. 256: C. A. Berenstein and M. A. Dostal, Analytically Uniform Spaces and their Applications to Convolution Equations. VII, 130 pages. 1972. DM 16,–

Vol. 257: R. B. Holmes, A Course on Optimization and Best Approximation. VIII, 233 pages. 1972. DM 20,–

Vol. 258: Séminaire de Probabilités VI. Edited by P. A. Meyer. VI, 253 pages. 1972. DM 22,–

Vol. 259: N. Moulis, Structures de Fredholm sur les Variétés Hilbertiennes. V, 123 pages. 1972. DM 16,–

Vol. 260: R. Godement and H. Jacquet, Zeta Functions of Simple Algebras. IX, 188 pages. 1972. DM 18,–

Vol. 261: A. Guichardet, Symmetric Hilbert Spaces and Related Topics. V, 197 pages. 1972. DM 18,–

Vol. 262: H. G. Zimmer, Computational Problems, Methods, and Results in Algebraic Number Theory. V, 103 pages. 1972. DM 16,–

Vol. 263: T. Parthasarathy, Selection Theorems and their Applications. VII, 101 pages. 1972. DM 16,–

Vol. 264: W. Messing, The Crystals Associated to Barsotti-Tate Groups: with Applications to Abelian Schemes. III, 190 pages. 1972. DM 18,–

Vol. 265: N. Saavedra Rivano, Catégories Tannakiennes. II, 418 pages. 1972. DM 26,–

Vol. 266: Conference on Harmonic Analysis. Edited by D. Gulick and R. L. Lipsman. VI, 323 pages. 1972. DM 24,–

Vol. 267: Numerische Lösung nichtlinearer partieller Differential- und Integro-Differentialgleichungen. Herausgegeben von R. Ansorge und W. Törnig. VI, 339 Seiten. 1972. DM 26,–

Vol. 268: C. G. Simader, On Dirichlet's Boundary Value Problem. IV, 238 pages. 1972. DM 20,–

Vol. 269: Théorie des Topos et Cohomologie Etale des Schémas. (SGA 4). Dirigé par M. Artin, A. Grothendieck et J. L. Verdier. XIX, 525 pages. 1972. DM 50,–

Vol. 270: Théorie des Topos et Cohomologie Etle des Schémas. Tome 2. (SGA 4). Dirige par M. Artin, A. Grothendieck et J. L. Verdier. V, 418 pages. 1972. DM 50,–

Vol. 271: J. P. May, The Geometry of Iterated Loop Spaces. IX, 175 pages. 1972. DM 18,–

Vol. 272: K. R. Parthasarathy and K. Schmidt, Positive Definite Kernels, Continuous Tensor Products, and Central Limit Theorems of Probability Theory. VI, 107 pages. 1972. DM 16,–

Vol. 273: U. Seip, Kompakt erzeugte Vektorräume und Analysis. IX, 119 Seiten. 1972. DM 16,–

Vol. 274: Toposes, Algebraic Geometry and Logic. Edited by. F. W. Lawvere. VI, 189 pages. 1972. DM 18,–

Vol. 275: Séminaire Pierre Lelong (Analyse) Année 1970–1971. VI, 181 pages. 1972. DM 18,–

Vol. 276: A. Borel, Représentations de Groupes Localement Compacts. V, 98 pages. 1972. DM 16,–

Vol. 277: Séminaire Banach. Edité par C. Houzel. VII, 229 pages. 1972. DM 20,–